Steampunk

Gear, Gadgets, and Gizmos

A Maker's Guide to Creating Modern Artifacts

About the Author

With an education in physics, Victorian history, and art, **Thomas Willeford**'s passion for inventing, art, and Steampunkery was probably inevitable. His company, Brute Force Studios, has been producing Steampunk products and spearheading the Steampunk movement since 1990. His work is an attempt to blur the line between art and engineering. If, upon viewing a piece, you do not ask, "Does that actually work?" Thomas considers the piece a failure. Sculpture and wearable art are his preferred forms, because he cannot draw or paint to save his life.

His creations are proudly worn by a veritable who's who of Steampunk elites worldwide, who count his work among their most prized possessions. At the time of publication, Thomas and his designs have been widely featured on television (MTV, BBC, ABC's *Castle* series), online (*Wired*, *Unplggd*, *Popular Mechanics*), and in print (*Marquis* magazine, *Bizarre* magazine, *Pirates* magazine, Art Donovan's *The Art of Steampunk*, Morgan Spurlock and Stan Lee's *Comic Con Episode IV: A Fan's Hope*, *True West* magazine). Thomas also contributed to the design of Alchemy Gothic's line of Steampunk jewelry and accessories. His artwork has been featured in numerous museum exhibits worldwide, including Penn State University's "STEAMpunk!," Dr Grymm's "Steampunk Bizarre," the Charles River Museum of Industry and Innovation's "Steampunk: Form and Function," and the Ashmolean Museum of the History of Science at Oxford's "Steampunk" (along with several others he can't quite remember, due to numerous lab accidents).

His alter ego, Lord Archibald "Feathers" Featherstone, has been showing his work throughout the world for some time now and hopes to continue supporting the cause of mad scientists everywhere for many years to come.

Thomas crafts items in his laboratory that bring to mind romance by gaslight, arcane science, the steam age, carnival sideshow curios, and aged materials from the vaults of Victorian England. Rusty things get him absolutely steamy. Mechanical clocks stop ticking as he passes (and pocket watches have been known to flee in terror) in the hopes that he will not notice them and then cannibalize them for parts. His curiosities have been hailed as imaginative oddities epitomizing the diverse landscape of Steampunk design.

Thomas currently resides in Harrisburg, Pennsylvania, with his beloved Lady Clankington and six spoiled cats.

Steampunk

Gear, Gadgets, and Gizmos

A Maker's Guide to Creating Modern Artifacts

Thomas Willeford

New York   Chicago   San Francisco
Lisbon   London   Madrid   Mexico City
Milan   New Delhi   San Juan
Seoul   Singapore   Sydney   Toronto

The McGraw-Hill Companies

Library of Congress Cataloging-in-Publication Data is on file.

McGraw-Hill books are available at special quantity discounts to use as premiums and sales promotions, or for use in corporate training programs. To contact a representative, please e-mail us at bulksales@mcgraw-hill.com.

Steampunk Gear, Gadgets, and Gizmos: A Maker's Guide to Creating Modern Artifacts

1 2 3 4 5 6 7 8 9 0   QDB QDB   10 9 8 7 6 5 4 3 2 1

ISBN 978-0-07-176236-6
MHID   0-07-176236-1

Sponsoring Editor Roger Stewart	**Indexer** Karin Arrigoni
Editorial Supervisor Janet Walden	**Production Supervisor** Jean Bodeaux
Project Manager Patricia Wallenburg	**Composition** Typewriting
Acquisitions Coordinator Joya Anthony	**Art Director, Cover** Jeff Weeks
Copy Editor Lisa Theobald	**Cover Designer** Jeff Weeks
Proofreader Paul Tyler	

Cover photography by Thomas Willeford; cover model, Sarah Herrick.

*This book is dedicated to My Grandparents,
Albert and Corrine McMullin. Thank you for exposing me
early in life to both Victorian splendor and Mad Science.*

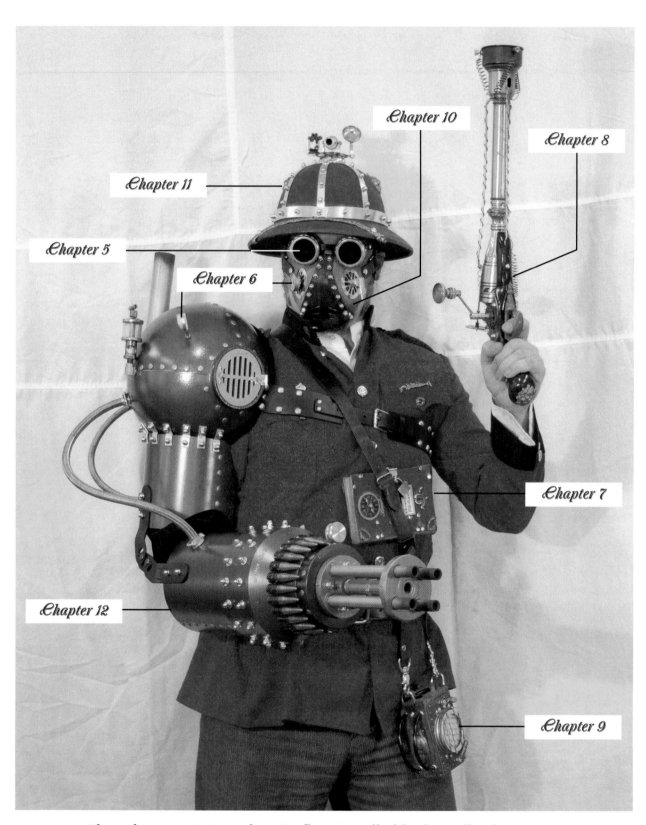

The Adventurer's Compleat Outfit as Detailed in the Following Pages

Contents

Part the First
Full Steam Ahead

Part the Second
Getting Steamed: Projects for Makers, Modders, and Assemblagers

Chapter 11

Voortman's Armoured Pith Helmet, from London's Finest Purveyor of Defensive and Deflective Haberdashery 163

Chapter 12

Professor Grimmelore's Mark I Superior Replacement Arm with Integrated Gatling Gun Attachment 177

Part the Third
Hastily Scribbled Laboratory Notes

A Letter of Introduction from a Noted Worthy

There was a time, a few years back, when Phil and I spent most of our time holed up in our studio writing *Girl Genius*. Well, we had small children, not a lot of money, and a web comic to put online three times a week. Right around that time, Steampunk as an aesthetic movement really started to gain momentum, and there was more and more wonderful art, gadgetry, and costuming online to distract me from my work. Not entirely a good thing, I'll admit, but it was fun.

One of the things I noticed early on was that everyone suddenly seemed to have gorgeous props, including the most wonderful brass goggles…and it seemed like the best of them were now coming from Brute Force Studios, the studio of that guy with the amazing, professional, movie-quality mechanical arm. We were seeing pictures of him all over the Internet, in his red coat and pith helmet.

You've probably seen them. They're those pictures that make you say, "Oh, my God. That is so cool. Hey, everyone around me who's trying to get work done, come over here and look at this amazing costume! I can't believe he made that himself!" Then all work stops as your co-workers gather around to check out the pictures and exclaim over the awesomeness. Great stuff—and everybody likes a good distraction from work, right?

Ah, but for some people, it doesn't end there. Some of us wind up thinking, "I wish *I* could make something like that." We start fantasizing about building our own wonderful props—maybe even our own fantastical Steampunk costumes capable of stopping work in offices and studios all over the country, nay, the world! All will marvel at our Glorious Steampunkery!

We plan our characters, imagine our adventures, even design our gadgetry in the margins of our school reports and office papers. *Our* stuff will be the stuff of *legend*. Yes, the daydreams are lovely.

Then the reality sets in—the reality of actually making these things. There is a world of difference between the concept sketch and the actual building of the thing, and many of us, raised in a culture where "leave it to the professionals" is too often casually thrown at amateur attempts at art, simply aren't sure where to start. Those of us involved with Steampunk are the lucky ones. Probably because the literary side of the genre tends to revere the archetype of the wild inventor, Steampunk naturally has a high percentage of makers: people who love to build clever devices and wonderful props, many of whom are frighteningly good at it, most of whom are willing to give advice and encouragement. And you know what? It turns out that being *able* to make things is closely tied to being willing to *try* to make things, and nothing makes us more willing to try something than seeing how it's done, and that it's possible.

We've seen that it's possible, so now, here's the guy who makes the stuff that *IS* the stuff of legend, to tell us how it's done.

Kaja Foglio, *Girl Genius*
www.girlgeniusonline.com

Muses and Mercenaries

With thanks to my long-suffering girlfriend, Sarah, for putting up with me as I became downright demented while writing this book.

To Mark Donnelly, for talking me down off the ledge and saying, "Don't panic, just throw the words at the page and we will clean it up later."

To Phil and Kaja Foglio, for their sheer inspirational Genius.

To Scott Church, for his generous photographic input.

To all of the members of Abney Park, for giving me something better to listen to than the whirring of gears in my head.

To Roger Stewart and the rest of the team at McGraw-Hill, for bringing this idea to fruition.

To my literary agent, Dr. Salkind, for bringing me the idea in the first place.

And to all of my other friends, family, and colleagues, who have been so supportive and encouraging through the process of writing this book.

In a Secret Lab Somewhere...

$\mathcal{W}$ELCOME. Welcome to an *anywhen* of Mad Scientists and daring Skyway Robbers, a *neverwhere* in which Airship Pirates play a deadly game of cat and mouse with Air Admirals of Her Majesty's Royal Aeronautical Corps; a Victorian(ish) Era that could have happened had history taken a different turn. Welcome, in short, to Steampunk. Imagine Verne hopping into his horseless carriage (perhaps like the one invented by Oliver Evans) and picking up Babbage along the way to an engagement. Wells and Tesla meet them at the door; inexplicably, they both had arrived much earlier than expected. Evidently, Edison's invitation was somehow lost in the post. They all meet at Arthur Conan Doyle's flat for tea and a chat. About an hour into it, Wells asks, "Does anyone have a pen? Shouldn't someone be writing this down?"

If you wish to visit this *anywhen* as more than a mere tourist, you might want to be suitably equipped. Every adventure worth the name has its dangers. There could be strange creatures, raging automatons, a severe lack of tea cakes—one must be prepared for these and other regrettable discomforts.

There are some incredible artists out there making ingenious devices to inspire the mind and please the eye. One of these artifacts can be yours with only the effort of disposing of a substantial bit of that excess income you have about you. I've seen a few big companies try and mass produce,

package, market, and sell "Steampunk." The results have been almost universally poorly made, aesthetically unpleasing (read Ugly), or more likely both. I want to be part of a third choice. I want to help you make your own artifacts. I can tell you from experience that there is nothing like the feeling you get when someone stares at you, open-mouthed and near speechless, as you inform him the piece he has been lusting over was made in your Secret Lair by your own hands. The parts are not expensive if you know where to look. I will show you those secret little places to get all those things that want to be made into other things.

I realize not everyone has a Death ray in his or her garage to cut those pesky complicated shapes, so here I will show you how to make all these projects using only the most common tools I could find. After reading this book, you should be able to buy all the parts to make your first pair of goggles and pay for this book for less than the price of buying a comparable set of goggles online, and you will still have the book to make more devices.

Other Mad Scientists might call me insane for giving away our most closely kept secrets, but here they are in this book. You do not need to have years of machining, engineering, metal working, and leather working experience to make these beautiful and slightly dangerous things. I have distilled those skills down to the bits you need to get the job done. I have done most of the skill-based heavy lifting so you can get on with the art.

I happen to be one of those artists I mentioned earlier, so you might be wondering why I would tell you how to make my designs that I worked so hard to develop? There are two reasons. Firstly, it will keep me from stagnating. If you are all out there making these wonderful things, I will need to come up with a whole new set of ideas to stay in business. Secondly, I like to think that after making some of the projects in this book, you will have a much greater appreciation of the craft and artistry that goes into creating this type of work.

I don't expect you simply to follow what I have laid out here as if it were just a set of blueprints; that's what minions do. You are made of much more adventurous stuff. You are going to make the basic pieces and then go wild combining other found objects and your own creations with the basic structures I show you here. Please expand the scope of what you find here. Add coils, switches, and old model parts. Make each piece your own invention.

This is not a book that should sit pristine on your coffee table or bookshelf. It should have greasy fingerprints all over it, many dog-eared pages, loads of notes scribbled in the margins, tears around the edges, maybe even a few scorched places. When you are done with this battered tome, and you hear the stomping of your nemesis's steam colossus coming toward your lair, you should have a veritable arsenal of Infernal Devices at your fingertips as you board your Radium Powered Aero-Dreadnaught and shout skyward, "You have messed with the wrong Mad Scientist!"

To my Fellow Scientific Enthusiasts, Mad or Even Just Mildly Annoyed

Now before any of you get your dander up and tell me how you have a fully developed character in your future-tech world, I do not consider what I've written to be *definitive*. (Although I believe an international law actually requires someone to go online and preach to the world about how wrong I am within 24 hours of the first copy of this book hitting the shelves.) They asked about my observations, and I complied. You might wonder, why me? Not to toot my own horn (well, at least not too loudly), but I'm pretty much a professional Steampunk. I got into it in the early 1980s when nobody even knew what to call it. With an education in physics, history (Victorian studies), and art, I was not qualified to do anything but be a professional Steampunk. What does one do with three relatively useless degrees? Make a costume company, of course! Corsets are our main product line, *but* we manage the occasional steam-powered replacement arm or personal ornithoptic flying machine...oh, and, of course, brass goggles.

Your Humble Servant,

Lord Archibald "Feathers" Featherstone
(aka Thomas Willeford)
www.bruteforcestudios.com

Part the First

Full Steam Ahead

Chapter 1

What Is Steampunk?

I see it as a reaction to the utter soullessness and disposability of modern tech. There are only so many garish space-eggs and tech bubbles you can look at before you just stop appreciating them. Steampunk harkens back to a time when technology was still novel and romantic, when the world was still marveling at its own cleverness with childlike pride and wonder, looking hopefully toward a strange and wonderful future.

—Richard Nagy, Datamancer.net

I've seen much debate online about what is and isn't *real Steampunk*. My least favorite is this one: "Steampunk is not real, so there are no rules and you can do anything you like and call it Steampunk." Sweeping definitions like this are not really helpful to the "goggle-curious." I personally have no fear of applying a bit of definition to help things along, so when asked in a recent interview to define Steampunk, I bravely said, "I'll write something up and have it to you next week."

So here's what I've come up with....

What Steampunk Is

According to subcultural folklore, the term "Steampunk" was first coined by author K. W. Jeter back in the early 1980s, during the heyday of cyberpunk. Instead of writing about cyberpunk's dystopian future, Steampunk authors such as Tim Powers, William Gibson, James Blaylock, and Bruce Sterling were looking back at a ruptured past, with technologies emerging out of sync with our timeline and societies struggling to cope. Although many people consider Jules Verne, H. G. Wells, and George Chetwynd Griffith (by the way, these are some of my favorite authors in the entire universe) to be Steampunk authors, I believe they are better categorized as science fiction writers of their time. These authors wrote amazing things about their own time or about a speculative future.

I am going to push the boat out here and risk taking a shot across the bow from the S.P.P.D. (SteamPunk Police Department). The first piece of media to come about that I would classify as truly Steampunk is the '60s television show *The Wild Wild West* (pilot episode, 1965). There, I said it. I looked and nothing else really fit the bill. Yes, *20,000 Leagues Under the Sea* came out in 1954, but it was a movie based on a novel published in 1869. *The Wild Wild West* had all my prerequisites for good honest Steampunk, without any stretching of the definition whatsoever. It had the gadgets, the alternative history, the mad scientists, and dashing heroes. Others might wish to stretch the definition, even if just to try and prove me wrong, but there is no question about whether *The Wild Wild West* was Steampunk or not—it just was.

According to that revered tome, the *Oxford English Dictionary*, Steampunk is defined as "a genre of science fiction that typically features steam-powered machinery, rather than advanced technology." And although the Oxford definition might be correct to a certain degree, I find it to be rather inadequate. Steampunk has graduated from a simple science fiction genre into a growing subculture. Its style is based on the clash of history and anachronism and infused with the demands and constraints of antiquated technology.

People are developing entire wardrobes and lifestyles based on where they park their airship. We used to joke that Steampunk is what happens when Goths discover the color brown, but that, too, would be an oversimplification. (Although I think the Steampunk world might owe some debt for its rapid growth and expansion to the fact that the success and popularity of Stephenie Meyer's Twilight Series vampire romance novels made vampires seem rather shiny and pathetic. This seems to have resulted in a mass migration of traditional Goths from the dark side to the sepia side—if one will permit the use of such a phrase.) And although Steampunk shares some elements with the gothic subculture at the fundamental level, such as their shared fascination with the strange and unusual, their foundation in literary works of the Victorian era, and their adoption of antiquated fashions of the late 19th century (though not exclusively, of course), Steampunk is still much more than that. Its literary roots make it a more character-driven world. One might dress Goth, but I seldom see people dressing generically Steampunk; instead, they put on their Lady Machinanna's best hunting outfit, holster their Dr. Visbaun's High Voltage Electro-Static Hand Cannon, and head off on a dinosaur hunt to deepest

far-flung Neptune. I hear it's Victoriasaurus season on the Baron's estate.

I discuss the two broad types of Steampunk worlds here: the alternative history, and retro-futurism/modernism.

Alternative History

It is the past but not quite the same as we remember it. Some shift has made science take an extra step, and society came along for the ride. Usually a few technologies or groups of technologies have made an advancement, while leaving other things behind. Maybe we can fly faster sooner, but Marconi's radio is nowhere in sight. This brand tends to be more character-driven. People often create highly developed *dramatis personae* to fit into their "brave new world."

Here's an example: It's 1879, and with the advent of Professor Grimmelore's Patented Helium Expander, mankind has taken to the skies like never before. Expanded helium has six times the lift of standard hydrogen and is so much safer. The sky is buzzing with airships kept from colliding only by the constant vigilance of the tireless operators of Her Majesty's Light Tower Semaphore Corps.

I can already hear moans from the crowd, but in all honesty, I believe the best example of this type of Steampunk, in the truest sense of the word, can be summed up in three words: *Wild Wild West*, the '60s television series I mentioned earlier. The show had it all: dashing heroes, mad science, plots to take over the world, and Victorian flare. Of course, my favorite flavor alternative history is Victorian/Western, but it can be most any time up to now where things went a bit…different. Imagine if the internal combustion engine was never developed and we had steam cars in the '40s and steam planes fighting WWII.

Retro-Futurism/Modernism

History, until the time of the story, tends to be the same as recorded. A group or culture has decided to pull its style from the past. Everything might have already changed, and now we are in the present day or the future, dealing with the consequences of what has already happened. Within the subculture this world is more of a "fashion/style statement" with not as much role-playing.

My favorite literary example of this is Neal Stephenson's *The Diamond Age*. It's the late 21st century and England has a new Queen Victoria. People are wearing full Victorian dress constructed of nearly indestructible and self-cleaning nano-fabrics. This is one of the few books I've read more than once. Another good example of this is a 2006 movie *Perfect Creature*. It featured an alternative history but brought it up to a present or near-present place and time.

The Steampunk aesthetic is not only about books, movies, and television, however. Fashion, music, and decor are all rapidly being integrated into this subculture. In mid-2006, with its album *Lost Horizons*, the band Abney Park became the first all-Steampunk band (as opposed to a band that occasionally plays goggle-friendly music). The band helped define the soundtrack of air piracy, and for a while, they were the only airship pirates. Other musicians soon followed—The Men That Will Not Be Blamed for Nothing, The Extraordinary Contraptions, Unextraordinary Gentlemen, Vernian Process, Dr. Steel, and Ghostfire, to name a few. A DJ could play all night at a party and not leave the Steampunk soundtrack.

What Steampunk Is Not ("Cog on a Stick")

> You know, just sticking a cog to your t-shirt does not make it a mechanical shirt.
>
> —Professor M. P. Donnelly

No one likes to say, "these are the Steampunk rules that you must follow," but there is a tendency within Steampunk art and fashion toward what I like to call "cog on a stick." The best way to avoid the "cog on a stick" effect is for things to at least have the illusion of functionality. You are allowed to do anything you like, but the rest of us are also allowed to point and laugh at you.

All of this might have a deeper significance. Why now? Steampunk has been around for almost 30 years, and some would say even longer. Why are more and more people suddenly re-embracing the old aesthetics? I have a theory: Steampunk is a rebellion. The "Chap Manifesto" (www.thechap.net/content/section_manifesto/index.html) calls it a "charmed revolution."

Look at our technology. Remember when radios and televisions were actually pieces of furniture encased in wooden cabinetry? They were pieces of art. Modern technology has none of that. The

---

Steampunk simply embodies a time and a place. The time…the late 19th century. The place…a steam-powered world, where air travel by fantastical dirigibles is as common as traveling by train or boat (or submarine). A place where national interests are vastly different than our own version of history. A place where the elegant and refined are as likely to get pulled into a grand adventure, as the workers, ruffians, and lower classes. A place where the idea of space travel is not so far-fetched. A place where lost civilizations are found and lost again. A place where anything is possible, and science can be twisted to meet one's own ends. That to me is the essence of Steampunk. It can have political overtones and commentary, or it can be straight escapist fiction. Either way, if it meets these criteria, it is Steampunk.

—Joshua A. Pfeiffer, aka Vernian Process

The word Steampunk refers to a particular genre, aesthetic and even a reality that "might have been." For some people, it's an evolved fantasy/reality that might have been had internal combustion engines never taken hold or even been invented. Steampunk for me is a reality that "aims to be rather than to seem." Indeed, it's an aesthetic that is heavily versed in a climate of invention and innovation. The construction and methods of operation, the kinetics of the piece are exposed and on the surface, as opposed to boxed in and hidden behind a false casing. The wonderful thing about a steam engine is that you can follow the path of power generation and function beginning with the fire box and boiler, follow the plumbing, valves, gauges, gears, d-valves, pistons, eccentric shafts, and fly-wheels all the way from the source of power to the final outcome of kinetic potential.

Within this architectural aesthetic, there are no false walls, drop ceilings, prefab decorative elements or the mundane presence of modern conveniences. Theatre is wonderful, but theatre is false in its constructions. There is nothing false or "out of the box" when talking about the Steampunk aesthetic. You'll find that there is an incredible complement between a variety of disparate materials that can usually be found in any Steampunk conceived of device…wood, brass, rivets, gears, lenses, cast iron, etc…. Steampunk is an honour to an era when people thought big, and worked hard to make things that last. It is not like the disposable culture of commodity that we have today. Care, artisanship, and craftsmanship were put into everything that was created.

—Sean Orlando, Kineticsteamworks.org

industrial revolution made more things available for the common person, which is great, but all of them were made the same, which is boring. Everything became so much about making money, and bland is now the common denominator. The first commercially available computers, for example, came in one color: putty. Attempts to change this on the corporate level have been limp, at best ("Now, in black!"). We have the technology, but where's the grace? Why aren't wood-grain outer casings available for our laptops? People would buy that. People on the street who have never even heard of Steampunk see my laptop bag and desperately ask me where they can buy one. When I tell them I made it, they offer to buy it right there. I would love to sell it to them, but then what would I use to hold my laptop?

The popularity of retro-styled cars is one indicator that people are hungry for more accessible beauty in their everyday lives. Steampunk looks to the past, where ornamentation was relished and encouraged, and it applies that desire for beauty *and* functionality to our modern lives. The Steampunk ethos and aesthetic makes it possible to apply modern technology with these old designs, while still being cost-effective.

Photo by Scott Church

Chapter 2

Tools of the Modern Mad Scientist

I am not going go lie to you and tell you, "You can make anything in this book with just a butter knife and a broken pair of tweezers." You are going to need tools. I am sympathetic to your plight. Not everyone has access to a CNC (that's machinist geek for Computer Numerical Control) milling machine or an industrial cutting laser.

All the projects in this book have been designed for every skill level, from the amateur hobbyist to the advanced machinist. You can construct the projects in your basement, spare room, or kitchen. On the other hand, if you do have access to a well-equipped workshop, that is fantastic, and I have no doubt you can find many different methods to complete the various stages of these projects. The more sophisticated the tools at your disposal (as well as your own skills and talents), the quicker and easier these various projects can be made.

I cannot assume, however, that every reader of this book has a complete industrial workshop at his or her disposal, so I have tried to explain these projects in the simplest possible way so that they can be completed with readily available (and surprisingly affordable) tools. Just because I am making an effort to explain these construction steps simply does not mean that more sophisticated techniques could not be used, however. Do not think, therefore, that you are locked into a certain methodology for making these devices. Feel free to use the patterns and designs and build them in whatever way seems most logical and expedient to you.

This is Steampunk. When it comes to construction, there are no hard and fast rules. A little time and ingenuity can more than compensate for the lack of advance tooling or equipment.

Tip

Bear in mind two primary axioms as we proceed: First, metal is always preferable to plastics. Second, screwed is always better than glued.

Must-Have Tools

Let us assume for the moment that you are new to the whole field of mad science and steampunkery, and you are just starting to assemble a selection of tools for your workshop or laboratory. Listed here are some tools that no mechanical mastermind should be without.

If you are new to all of this, please do not be intimidated by the following tool lists, and if you are still at an early stage in your tool acquisition, I suggest that you look through these projects and examine the necessary tool lists. I recommend that you begin with projects for which you already have most of the tools to make. Do not try to rush out and buy loads and loads of tools to complete everything in this book (or to begin work on an automated steam-powered colossus). Your workshop will grow and evolve over time as you find more and more useful tools.

Hand Tools

Almost everything in this book can be made with simple hand tools. Bear in mind that, in fact, most things constructed in the 19th century were made exclusively with simple hand tools. Yes, of course they did have power tools, but nothing like the portable electric power tools so commonplace today. You'll find the following hand tools useful:

- A selection of screwdrivers in a variety of sizes (both Phillips and flat)
- Needle-nose pliers: two pair minimum
- Larger strong pliers (such as lineman's pliers)
- Hacksaw
- Coping saw
- Hammers in at least three different sizes (weights): If one of them is a 2 lb. brass mallet, all the better
- Good-quality utility knife with replaceable blades
- Metal files
- Sanding block (with a variety of different grit sandpapers)
- A variety of clamps in different sizes
- Diagonal cutters or wire cutters (dykes)
- Wire strippers
- Rotary hole punch (for leather)
- Heavy duty scissors (for cutting leather and such)
- Bench vise
- Spring-loaded center punch
- Straight edge
- Small increment measuring tape or ruler
- Fine point magic markers
- Aviator snips (or other metal shears)

Handheld Power Tools

Power tools can make the work go faster, but they aren't always necessary to create the projects presented in this book. When you're drilling a simple hole for one of your projects, you can use a readily available cordless power drill, but you could also choose the Victorian option and use a traditional brace and bit. Although I frequently use an industrial cutting laser in making my projects, everything in this book could be cut simply using a traditional hand saw or utility knife.

These power tools, however, can help the work go faster:

- Electric power drill (preferably cordless) with an assortment of drill bits
- Electric rotary power tool (such as a Dremel): the variety of interchangeable bits for these make them a virtual necessity in any workshop
- Jigsaw, saber saw, or other reciprocating power hand saw

Optional but Recommended Tools

Although the following tools are not required to make the projects in this book, they certainly make the process easier:

∽ Small pipe bender

∽ Tap and die set

∽ Belt grinder

∽ Belt sander

∽ Drill press, which can also be used as a lathe

∽ Band saw with fine-tooth metal cutting blade

∽ Computer and printer (preferably color) for patterns and gauges and stuff

∽ Set of small X-ACTO knives

∽ Soldering torch and supplies

∽ Dial or digital caliper

∽ Small (5 lb.) anvil

Ridiculous Tools

These tools are certainly not necessary but oh so bloody useful:

∽ Larger (75 lb.) anvil

∽ Bookbinder's vise

∽ CNC industrial milling machine

∽ 3D printer

∽ 50 watt Kerns 24-by-48–inch CNC cutting laser

∽ Metal lathe (CNC if possible)

∽ Sheet metal break

∽ 10-ton pneumatic press

∽ Army of minions

Finding Tools

Secondhand hand tools and power tools are readily available from flea markets and charity shops as well as online through auction services (such as eBay) or classified listings (such as Craigslist). If you want to buy new tools, you should be able to pick up cheap but serviceable versions of everything on the hand tool list for under $50 (US).

When shopping for power tools, be certain to do some comparison shopping. Without mentioning any specific company names, suffice it to say that convenience comes with a price. Discount tool suppliers can be found throughout the country (most other western countries have them too). Do some research and watch for clearance sales. When comparing tools and prices, ask yourself, How frequently is it going to be used? If you are intending to use the tool for 50 or 100 hours a week, it might be worth investing in a better quality tool that is up to the task. If, on the other hand, you intend to use your tools only occasionally for a few projects, cheaper (entry level) hand and power tools should accomplish the job adequately.

As you scour your local haunts for all of the other parts necessary to any good Steampunk laboratory (see Chapter 3), keep a sharp eye out for useful tools as well. Let's face it, building a device that purports to be from the 19th century, using nothing but 19th century tools would be incredibly awesome (even if slightly challenging).

Chapter 3

The Art and Philosophy of Scavenging

Where do you get that stuff? This is perhaps the second question I'm often asked after I explain what a piece used to be before I brutally beat it into its present shape. The first question is usually, *Where did you get the idea for that?* I keep my eyes open for parts all the time. Most of the time I hunt down this stuff myself, but the habit has infected many of my friends, too. They'll call me up and say, "I saw this thing at a yard sale and thought of you. By the way, you owe me ten dollars for it." Sometimes I take the widget just to be polite, but sometimes I shove the money into their hand and yell, "It's mine now! You can't have it back!" as I run back the lab laughing maniacally.

There is a sort of art and philosophy to the scavenging process. It is impossible, of course, to teach someone (anyone) to think creatively. But by virtue of the fact that you are reading this book, I have to conclude that you have at least some degree of creative impulse. This is good, because it will make the entire process not only easier, but fundamentally possible.

Antique and Flea Market Finds

I scour local flea markets to find most of the oddball materials I use, such as clock parts, antique mechanical devices, house hardware, boxes of springs, and the like. I am fortunate, here in Pennsylvania, that we have the second oldest population in the country and that I live in a relatively rural area. The older folks in my neighborhood are very good at collecting and hoarding, and 19th century antiques are in abundance in this area. This is not to say that such things would not be equally readily available in other parts of the country—I can only speak for myself here.

I am such a frequent visitor to the flea markets in the area that the regular dealers have come to know me quite well, and they keep their eyes peeled for things they think I might like or be able to use. And even better, they frequently reserve special tasty tidbits just for me. The downside, however, is that they might charge me slightly more for an item than they would charge someone else, if they think I am a certainty as a buyer. This is a balancing act, and I cultivate those relationships carefully.

Tip

It can often be beneficial for you to take the time to listen to a seller's seemingly irrelevant stories. This helps cultivate a relationship with these sellers and sometimes those stories are not so irrelevant.

Older populations are also pretty good at something else—namely, moving to warmer climates—and when they do, they frequently offload a lifetime of collected junk—umm…I mean antique treasures. And this brings me to the subject of working with antiques (no, not the elderly—the flea market finds!).

It might sound strange to say, but as a general rule, when I'm looking for antiques, I tend to *avoid antique shop*s. Chances are, antique dealers have already done their research and are charging appropriate retail for all their wares. Not to mention the fact that if an antique dealer had any idea what I intended to do with that carefully collected, restored, and displayed antique, he or she might have a stroke.

When Old Means Gold (and When It Doesn't)

Antiques, as you know, can have great value in their own right. Before the dawn of the disposable consumer goods that fill and clutter our modern lives, everyday items were designed to last generations. At this point, many well-constructed items might not have intrinsic value in their own right—which is to say that they might be constructed from simple materials such as wood or cast iron instead of precious metals or stones, and they might be unusable for their intended purposes. And they might not even have any outright artistic value that would make them costly or desirable. But their longevity as utilitarian objects might mean that they are exceedingly rare, and therefore, quite valuable. Although the industrial revolution of the 19th century spawned the age of mass production, some of these items might or might not have been manufactured in large numbers. Therefore, you must proceed carefully when working with antiques. Do your research *before* you tear into it to harvest parts.

Tip

Internet-capable smart phones can prove invaluable for checking out an item's value while you are actually shopping, before you even purchase the item; but do be discreet about it.

I learned this lesson the hard way. Hopefully you can learn from my mistake. I once bought an old clock at a flea market for $10 or $20, because it contained several gears, wheels, and springs, which I thought I could use for other projects. I proceeded to tear my way into it, salvaging all the yummy brass bits I was after. Only later did I learn that it was not a late 19th century, mass-produced piece as I thought, but a handmade clock of the mid-18th century, and the internal mechanism I had just destroyed was worth about $2500 (not to mention the valuable handmade wooden case and body, which I also destroyed in the process). It was a lesson hard learned, and I am not proud of this. Now whenever I return with a pile of booty from the flea market, I immediately sit down with the Internet to get a sense of the value of my purchases. Do I really want to cut into that first edition *Mark Twain* to turn it into a portable hard drive (see Chapter 7)? Probably not.

Conversely, "old" does not necessarily equate to "valuable." An ignition coil for a Model T Ford (see Chapter 9), for example, was a semi-disposable item when it was manufactured. Millions of these were manufactured, sold, and used. They are the Edwardian equivalent to today's dead Duracell. That's not to say, however, that they are without value entirely. I like to reuse the housing and the terminals for T-pods (Chapter 9), even after their lifespan as a usable power source has long since expired.

In some cases, items were simply inexpensive even when new. Consider, for example, an early 20th century egg beater. Every kitchen had one. Today, they look cool, with interesting gearing and crank handles. They definitely offer some usable parts. But should I be worried about destroying an old egg beater and reusing its parts in a new contraption, simply because it is old? I do not think so. It was never valuable and will not likely ever be valuable. There are few in the 21st century who are searching for old egg beaters to use on a daily basis. But these items were made to last, and you'll find plenty of them in junk stores and flea markets, and even in antique stores. I feel perfectly justified repurposing usable parts from them. Well, I did before writing this book; then my editor pointed me toward www.misterfindit.com/egg-beaters.html. I had no idea people collected them. See what I mean about researching these things? Like the site says, most egg beaters are not worth much, and the same goes for alarm clocks, old cameras, and typewriters. I remember being able to go to flea markets and get old adding machines free at the end of the day because the seller did not wish to pick it up.

Books can be tricky. Many people have an almost religious devotion to collecting and saving old books, which is something I can fully appreciate. But like anything else, just because a book is old doesn't make it necessarily valuable or precious. Fortunately, books are one of the easiest things to check out value-wise, and with a simple Internet search of title, author, date of publication, and so on, you can learn a lot about what you have. Is this book the extremely rare 1897 edition with the original yellow cloth cover? Or is it, in fact, the 1932, eighth edition, worth a few dollars or a few pennies?

Note

If you want to photograph an item and research it later, be sure to ask the current owner (the seller) before snapping the shot. Most sellers don't mind you taking a photo, but some do.

Even rare books, if significantly damaged, might lack sufficient value to be considered precious. But even damaged old books sometimes contain plates, engravings, or maps that are worth salvaging. As with all antiques, condition is everything. (Just watch the PBS series *Antiques Roadshow* if you don't believe me.) Unfixable clocks or pocket watches frequently have little or no value. If in working condition, however, they can fetch large sums. You could, of course, look into the process of restoring the antiques you find in hopes of rejuvenating them to restore some of their lost value. If that is a road you choose to walk, good for you. I have a lot of respect for antiques restorers, but the process of antique restoration is complicated and beyond the scope of this book.

In short, when it comes to antiques, if you are not certain what you have or what it might be worth, put down the hammer and do a little research first.

Finding Parts and Supplies That Are Not Antique

I am often asked, *Where do you get all your leather?* And the snarky part of me replies, "At the leather store—where else?" Now I fully appreciate that all readers won't have a leather dealer within a mile or two of their home. But, then again, most readers are also not likely to go through several full hides a week. One must be adaptable and willing to improvise.

To create the projects contained within this book, most readers should be able to source a significant supply of leather by recycling old leather coats, bags, belts, purses, or luggage—all of which are often available at a local charity shop or yard sales. It's always worth knowing when local charity shops have their weekly or monthly sales. Get out there and start a stockpile of materials. It will pay off. I have no way of knowing what you will find out there, but, then again, neither do you. Go out, have a good look, and think creatively. What shapes do you see? Can you find leather of appropriate weights, thicknesses, or colors? The answers to these questions are entirely dependent, of course, on the nature of your project.

Brass items are also indispensable for repurposing into Steampunk gear. While you're browsing at a shop, flea market, or yard sale, keep your eyes open for "the color" (thanks to the TV show *Deadwood* for this term). After years of practice, I can spot a glint of brass from 30 yards away and virtually smell it inside a closed box under the next table. In searching for your brassy bits, keeps your eyes open for old lamps, candlesticks, random hardware, fire screens, and the like. Note that old brass can be hiding beneath layers of tarnish and patina. Don't be fooled by the lack of shine, because these often offer the tastiest bits for reuse.

Brass and other metal odds and ends, of course, do not necessarily need to be old, nor do they need to have begun life as something else to be useful for Steampunk purposes. Spend an hour or two at a hardware store, and take a look at everything in the hardware section. Look at all of the metal plumbing parts. Look through all the lighting parts. Familiarize yourself with bits and parts and pieces. Inspiration should follow as night follows the day.

Of course, you'll be shopping at hardware stores for many of the necessaries of your projects, such as machine screws, bolts, nuts, washers, couplings, and so on. But don't limit yourself to the required items only. Could those sink drain covers be vents on an altitude mask? Does that lamp finial want to be part of your ray-gun blaster? Could that plastic sphere in the garden center be used for a space helmet or part of a diving bell? Cast a creative eye about and you are likely to conceive of dozens of projects that you probably don't even have time to complete.

Hobby shops and craft centers can also be great sources for supplies and materials, especially for paints and varnishes. Look through the craft kits,

Beware the Weekend Warriors

A word of caution about hobby shops, craft stores, and hardware centers is warranted here. If at all humanly possible, avoid these locations on weekends and holidays at all costs. To ignore this advice is to proceed at your own peril. Many intended simple trips to pick up some paint or brass bolts can turn into epic treks of Burroughs-ian proportions, as you find yourself swimming through hordes of impatient and cranky children, whose keepers fight over discounted yarn or grass seed. This can, on occasion, inspire a violent rampage. If this occurs, don't say you weren't forewarned. Then hurry home to construct a six-story-tall giant killer robot. If you must rampage, you might as well do so in inimitable Steampunk style.

bottles, and frames. Why not check out jewelry and scrapbooking supply stores for ornaments and embellishments? Metal embossing supply stores can offer multipurpose sheets of brass, copper, or pewter. You'll find many things that want to be made into other things.

You can also explore hobby stores that sell parts and supplies for remote control cars, boats, and planes. They are frequently a reliable source for brass, copper, and aluminum tubing, sheet, and rods. And while you're there, check out the model train set supplies. Miniature lamps, boilers, switches, ducting, servos, and other small items can easily become parts of other projects. Does that O scale tanker car want to be part of the respirator for your diving bell? You decide.

Knowing What to Get

At least as far as the projects in this book are concerned, I will tell you exactly what is required for each project and what sort of alternatives might be suitable. After you have made several of the projects in this book, the process should be much clearer. If you already understand how to look for gold in unlikely places, you have likely skipped this chapter entirely and jumped straight to one of the project chapters—which means, of course, that you are not actually reading these words. If, however, you are still following along, here are a few tips that might prove helpful when you're hunting components.

If Possible, Avoid Plastic

Painted plastics, even when well done, always tend to look like plastic. And plastic is a hallmark of contemporary design and not at all Victorian-esque. If you must use plastic in the construction of your piece, do your very best to disguise it with more period-appropriate materials. See, for example, the way we constructed the metal bezel to hide the plastic housing on the plasma light in the T-pod in Chapter 9.

Painting a pair of plastic welding goggles to make them look cool is fine, as far as that goes, but it is a process of disguise, not of construction. This is a book about making, not disguising. Many steampunks will start out with plastic parts and accessories and then upgrade to devices and apparatuses of more robust construction as they become more serious.

There are of course exceptions to all of this. The Superior Replacement Arm in Chapter 12 is almost completely made from plastic. You will have to read on to see how we deal with that.

Know What You Are Getting

This point goes beyond purchasing antiques. Consider, for example, that brass lamp you just found in the charity shop. Is it actually made of brass? Or is it cast zinc or pot metal with a thin brass electroplate? Does it matter? Well, it might matter, depending on how you intend to use it.

Brass is soft and easy to cut. Steel and other alloys tend not to be soft and are, therefore, not as easy to work with. Furthermore, if a piece is plated brass, and you cut off an end or a portion of something that is brass plated, you'll have to contend with the raw (unplated) edge where you made your cut.

Nevertheless, mixed metals are perfectly acceptable for use in Steampunk creations—in fact, not only are they acceptable, but they look pretty spiffy when used in combination. As a case in point, check out the Alchemy Empire: Steampunk line of jewelry and accessories produced by Alchemy Gothic in England (www.onlinegothic.co.uk). It doesn't all have to be brass—it's just that brass and cast iron are the two most common metals used throughout the 19th century.

Get Familiar with Victorian Design and Aesthetic

For a Steampunk design to be truly successful, it should, at least to some degree, look like it could have been constructed in a *neverwhen* of the mid- to late 19th century. By World War I (1914–18), the aesthetic, technology, and the entire world were irrevocably changed forever.

Consider, for example, the classic design of H. G. Wells's time machine. Had he written his 1895 book, *The Time Machine*, 40 years later, the actual machine might have been described as a rocket ship, taken straight out of a Flash Gordon comic book, or perhaps something like a big glass sphere (as apparent at the 1939 World's Fair). But instead, Wells wrote a Victorian-era science fiction novel. As such, the technology in the book is derived from the aesthetic of his time. The time machine is built from a velvet upholstered salon-style armchair, adorned with polished wood and shiny brass. It is opulent, sumptuous, and solidly built.

Like Wells's time machine, your wondrous contrivances should be beautiful and aesthetically pleasing, as well as intriguing and fascinating. When you are designing your own Steampunk contraptions, bear the Victorian aesthetic in mind. No matter what you are making, try to select the best possible materials at your disposal. When in doubt, err on the side of opulence and splendor.

Learn to See Something Else

This might be the trickiest part of the process to explain. I think it is virtually impossible to explain to someone how to look at objects creatively—that is, as something that they're not. If you were the type of child who turned every cardboard box into a rocket ship, tank, or fort, simply because it was a suitable enclosure, this process might be easier for you to understand.

As you examine objects, notice their shapes and evaluate their proportions. When I see a 1920s hand pump fire extinguisher, I might turn it on its side and envision its movement as a large piston or ram. Or, in the curves at the base of a bulb-shaped lamp, I might see a place to house the radium chamber for an electrostatic hand cannon (see Chapter 8). When I initially started making brass goggles back in 1998, it was because I was in the hardware store picking up some plumbing parts for home repair; I was inspired by the shape of the 1 1/2 inch brass slip joints (see Chapter 5). They said "goggles."

Tip

Sometimes I'll start a project and let the contraption evolve by allowing the huge stockpile of parts and pieces to direct me in the final design. If you are comfortable with the process and want to attempt this approach in building your gear, then so much the better.

To see creatively, you need to cultivate a critical eye. Envision in your creative mind how the final piece is going to look—not in every detail, but get as close as you can. Think about your imagined design from all angles in your mind's eye, and figure out how all the components relate to each other. This gives you a great sense of the parts you need. Then go out and find those shapes. Of course, you aren't likely to find items of exactly those shapes you've imagined—it's not like you can pop over to Steampunks-R-Us and buy exactly what you want. So you might need to decide whether you can construct the pieces you are missing (which brings us back to tools, talent, and experience) or if you are going to adapt your design to suit the components that you have been able to locate.

The more projects you start and complete, the easier the process will become. Then again, if my experience is typical of most steampunks, as the years pass (dare I say decades?), your projects will indeed become easier to design, but your designs are likely to grow increasing difficult and complicated as you get more creative.

If, try as you might, you find yourself incapable of critical shape analysis or of creative adaptation of existing components, never fear. That's where this book can help. Simply choose a project from the following chapters and proceed along step-by-step. The design work has already been done for you; you have only to collect and build.

Have Fun

In the end, this is all supposed to be fun. There is no such thing as failure in the design and construction of Steampunk artifacts. If it doesn't look exactly like the pictures, who cares? You will build skills as you build each new project. Apply what you have learned to the next project.

On the other hand, if your project is a success, you will undoubtedly display or carry your creation for years to come, beaming with pride every time someone says, "Oh wow! That is amazing. Where on Earth did you get it?"

Chapter 4

Gear Mining—Or, How to Dissect a Cuckoo

So there I was, at the Bursledon car boot sale (that's the British equivalent of a flea market for our colonial readers) in Hampshire, England, with my friend James looking for raw materials from which to make things. And there she was, the archetype of the "little old lady," sitting behind her table, selling a few bits of old stuff. Right in the middle of the table is this perfect, though slightly damaged, cuckoo clock. I distracted James by pointing toward another table and saying, "What is that brass thing over there?" My hand shot out like it was spring loaded and I grabbed the clock before he could notice.

"How much for the old, broken clock?" I said, trying to down sell the piece and keep any hint of excitement from my voice. Little old ladies can smell desperation.

She took it from my hands and said, "My granddad brought this back after the war. Nobody seems to want to fix them anymore, but I'm asking a fiver for it." Only five pounds? I was trying not to shake. I could see that James had figured out I had "clock blocked" him by the look on his face. I handed her the five pounds just as our friend Amanda approached.

"Hey did you find anything? All I found was some creepy doll heads and an old bench vice. Hey! Did you just get that clock?" she asked, somewhat put out that she had not found it first.

"Yes! And I bet it has some great parts inside." I said, a bit too loud and enthusiastically, as I

proceeded to tear the case apart right there with my bare hands—hardly noticing the gasp of the little old lady as she clutched her chest in shock and horror. Everyone in the area had gone rather quiet.

"Tom, mate, time to go. I see pitchforks and torches in your immediate future." James took my arm to lead me away.

Please learn from this cautionary tale and from my momentary lapse of consideration before proclaiming your intentions of pillaging a family heirloom for parts.

Project Description

Unlike the projects that follow in this book, this chapter is not about creation but, rather, about destruction (although some might go so far as to say desecration). Well, no, we aren't actually destroying anything—more like recycling, or upcycling.

Many of the flea markets and estate sales I go to have a cuckoo clock or two for sale, and in my opinion, cuckoo clocks generally have the best gears. Some of them might be valuable Victorian antiques, but most of the ones I've found over the years were cheaply mass-produced between the 1940s and 1970s as either souvenirs or as exports from Switzerland, Germany, Austria, or even Japan for home decor. I research each one carefully before subjecting it to the tortures of the operating table or workbench. It is usually missing some crucial part or has been damaged in some way, and after researching it, I frequently find that it will cost more to restore it than it would actually be worth in good working condition. At this point its fate is sealed.

Note

If you feel preciously about vintage or antique clocks and think that doing anything with them beyond making them tick and keep perfect time is somehow sacrilegious, then please look away now or skip to the next chapter. This is, after all, a book on Steampunk projects, not on clock repair.

Materials

This project requires one dead clock. Meet our gear donor, Ken:

Well, more precisely, this is a Mi Ken cuckoo clock, made in Japan sometime between 1930–69. A quick search at the time of writing shows this make and model is currently available secondhand (damaged, of course) online for $10 to $20 US. I

feel no compunction or hesitation whatsoever about pillaging Ken for parts, despite the occasional mournful plea issuing from one of the cuckoo whistles.

I have no doubt that rather than slowly deteriorating as a rotting clock corpse, Ken will enjoy an exciting and varied afterlife after he has been reincarnated as all sorts of wondrous gadgets and gizmos.

Tools

Although all of the tools listed here (and shown) might not be necessary, most of these tools will be of use for this project:

- Rotary tool with cut-off wheel
- Small adjustable wrench

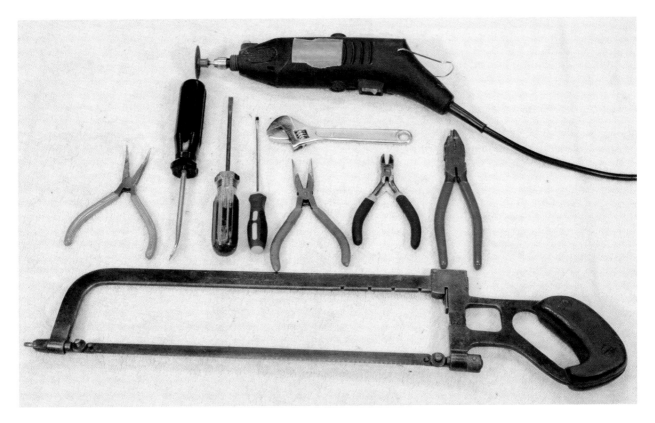

∽ Angled needle-nose pliers

∽ Crow foot pry tool

∽ Medium-size flathead screwdriver

∽ Small flathead screwdriver

∽ Needle-nose pliers

∽ Small diagonal cutters

∽ Medium-size pliers

∽ Hacksaw

　Also very useful to have are the following:

∽ Towel

∽ Safety glasses

Dissection and Autopsy

Let us assume, for the remainder of this chapter, that you are sitting in front of Ken's twin brother (as well as this book, of course). The following entire dissection process should take about 20 minutes or so (depending, of course, on any difficulties you might encounter).

Step 1　If your clock still has its drop weights, remove them first. (Over the years, my stockpile of these pinecone-shaped weights has grown rather ponderous. I keep them around with the intent of reusing them for something, but as yet I have not come up with a good enough idea. If you come across any inspiration in this regard, do let me know.)

　Once you have removed the weights (if it has any), you can remove any weight chains attached to your cadaver...umm, I mean clock, of course.

Each chain should have a large ring on its length somewhere, which is designed as an anchor or stop to prevent the chain going all the way up into the clock mechanism.

Simply cut off or remove this ring from the chain and then pull on the other end of the chain to draw its full length through and out of the clock.

Step 2 Lay the clock body on an old towel on your work surface. Although it is possible that your clock could contain parts coated in grease or oil (depending on the type and age of your gear donor), that's not why the towel is necessary. As you begin the pillaging process, you'll want to salvage any and all useful parts your donor has to offer. Experience has shown that small little bits such as nuts, gears, and springs have a tendency to bounce and roll off of smooth, hard work surfaces. The towel, therefore, works like a sort of net to catch and hold all of these (frequently vital) bits.

Step 3 Remove the push rods that are connected to the bellows. The bellows operate the whistles, which in turn give the cuckoo its voice. A small pair of diagonal cutters should suffice for this purpose.

Step 4 Using a small flathead screwdriver, on the side of the clock housing, unscrew the screws that hold the bellows and whistles in place.

Note

It is worth mentioning at this point that you don't want to throw anything away. Now I appreciate that you probably don't have much of a plan involving cuckoo bellows at this point, but that doesn't mean that your next project won't need them.

Step 5 Turn the clock over (face up). You should see a small nut that has been finger-tightened in place to secure the hands. Unscrew it and remove the hands.

Step 6 Turn the clock over once again (face down). You'll see four screws attaching the mechanism to the interior of the face of the clock. Remove these four screws and lift the mechanism out of the clock body.

Tip

You now have two primary parts: the metal geared clock mechanism and the carved wooden clock carcass. Consider turning the clock carcass into a birdhouse (not covered in this book) so that real birds can go in and out of the little door in place of that strange little plastic timekeeper. I have also been known on occasion to turn a clock body into a stylized backpack of sorts. Use your imagination and see if you can find a new purpose and new life for each of Ken's components.

Step 7　Using a pair of needle-nose pliers, unbend and remove any spare push rods.

Step 8　You should see two small screws securing a bracket that supports the pendulum hanger. Unscrew them and remove the hanger.

Step 9　Looking at the front of the mechanism, with the birdy on top, you should see two small screws on the right-hand side that secure two small brackets that support the cuckoo bird (see the preceding image). Unscrew them and remove the birdy assembly.

Step 10　Turn over the mechanism and examine the back side. Locate the nuts on all four corners of the back plate threaded on to the ends of the pillars. Use your wrench to unscrew those nuts.

Step 11　Examine the end of the central shaft on the back panel. Look closely and you should see a small wire pin holding it in place. Remove that pin.

Step 12　Now it's time to peel the mechanism open like an oyster. When you do this, be sure that you do so directly over the center of your towel, because geary, coggy, brassy goodies are likely to drop out of it like a busted piñata.

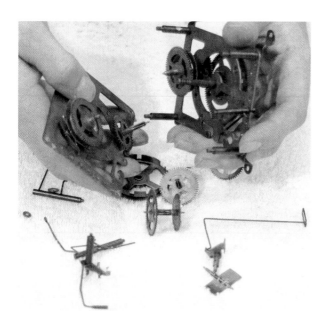

If that fails, there will be no other recourse than to amputate.

Step 13 Pull away all the pieces that you can separate using mild amounts of force. If you encounter any uncooperative gears, examine the assembly for any more of those pesky little wire pins. If you find any, remove them with your needle-nose pliers.

Step 14 Revisit any stubborn gears. You might find that a gear or two will cling stubbornly (either through sheer determination, or perhaps just out of habit) to its foothold on the body plate. To free a gear like this from your plate, try a crow foot pry tool (which often works).

Step 15 If you intend to use the front plate for anything, you might want to remove the four pillars and the bracket feet. To do this, you will likely need to cut or grind them off.

After you have separated your clock corpse into all available cogs, gears, pins, pillars, chains, and so on, you should have a pile of parts that resembles what's shown below.

Now, a moment of silence for Ken.

If you really are Steampunk, looking at this pile of parts should inspire all sorts of creative ideas and projects. If you cannot look at these bits of mechanical marvel and get excited about making things, perhaps you should divert your energies elsewhere. I suggest macramé.

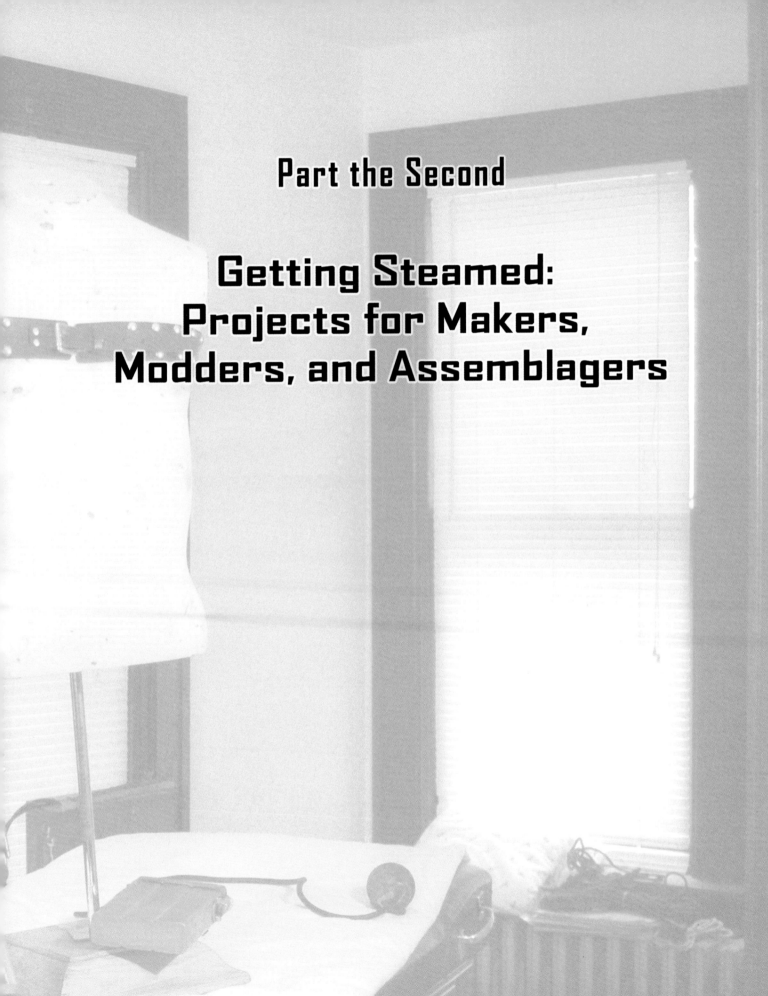

Part the Second

Getting Steamed:
Projects for Makers,
Modders, and Assemblagers

Chapter 5

Aetheric Ray Deflector
Solid Brass Goggles

"Alright, ladies!" the Sergeant yelled to the small group of six men and two women. "As we pass o'r the *A.S.V. Orion*, you sorry lot will jump out this 'ere 'atch and land on top 'er gas bag. Then Corporal Grayson 'ere will cut through the skin and you will board 'er. We will come 'round an' pass under 'er exactly six minutes after you leave. Everyone got that? Only one pass. You don't make it, you get to enjoy the pleasure of the Count's fine 'ospitality till 'e sees fit to toss your bum out o'er the channel."

Sergeant Armstrong always gave the most encouraging speeches just before every boarding action. He once told a midshipman he was going to need "further training." This did not concern the middy much till it was explained that it involved Sergeant Armstrong throwing him in front of an oncoming train if he failed to get it right the next time.

"I am certain none of you 'ave any questions, so line up be'ind the Corporal 'ere. Goggles down!... On my mark.... GO!"

—From *My Life in the Air*, by Ms. Adelaide Grayson
(Major, Her Majesty's Aero-Forces, Ret.)

hether you are a dashing airship pirate (or "privateer" if you prefer to feign an air of legitimacy), skywayman (not to be confused with the more mundane "highwayman"), or simply the maddest of scientists, nothing screams STEAMPUNK! quite as loudly as a good pair of genuine brass goggles. Certainly, you could buy some cheap plastic welding goggles and paint them up, but then you would actually have to be seen wearing them.

Once, when I was walking about London with a pair of my handmade goggles hanging about my neck, a woman darted out of a shop to ask me if I was a steampunk. And, of course, she wanted to

know where she might get a pair of her own. Stylish brass goggles are the designer shoes of the neo-Victorian world, and quality speaks for itself.

Project Description

In this chapter, we'll build a simple and nearly indestructible set of Steampunk brass goggles. I began manufacturing these in 1998, and this design has been frequently copied by others who could not come up with their own designs—but my design is rarely equaled. When I first started producing these goggles, the public didn't quite know what to make of them. "Why would you want goggles made of brass?" I was frequently asked. "Well it is a sort of Steampunk aesthetic," I would reply. Then the "What is Steampunk?" questions would inevitably follow, and I would try to explain, but I would invariably end up saying something to the extent of "Well just wait…. You'll see…. It will come around." And so it has. The fact that you are reading this book, and that you are specifically reading this chapter, because you are interested in making for yourself a genuine pair of brass goggles, is evidence that this design has finally become a standard.

What You'll Need

Now then, let's get started. To complete this project, you will need to get your hands on the following materials and tools.

Materials

- 1, 1 1/2-inch, 22-gauge, rough brass slip coupling (pipe fitting), at least 4 inches long, with nuts
- 1 roll masking tape, 1 1/2 inches wide
- 1 can flat black spray paint
- 1 piece of soft leather, light jacket weight, at least 6-by-10 inches total
- 18-inch (approx.) length of 1/4-inch cord
- 20 (approx.) small brass double-cap rivets
- 2 feet (approx.) of light belt-weight leather, at least 1-inch wide
- 10 (approx.) medium-size brass double-cap rivets
- 5 (approx.) large brass double-cap rivets
- 3, 1/2-inch "D" rings
- 1, 1/2-inch square or oval center bar buckle
- 1/8-inch transparent acrylic sheet, at least 4-by-3 inches

You can acquire the slip coupling at your local plumbing supply store. The cheapest way to get both the jacket- and the belt-weight leathers is to go to a thrift store and buy an old leather jacket and a lightweight leather belt; otherwise, you can get them from a leather craft store along with the buckle, the rivets, and the "D" rings. The hardware can also be purchased at a craft store or sizable fabric store, preferably one with a craft department. As for the acrylic sheet, try asking at a place that offers replacement glass and window repair. If you tell them what you are making, they might even just give you a small piece large enough for your needs, so you can avoid having to buy a much larger piece. You can usually find the spray paint, tape, glue, and the acrylic sheets at a hardware or builder supply store.

Tools

- Paper
- Scissors
- Rotary punch
- Marker pen
- Leather gloves
- Center punch
- Safety glasses
- Bench vise
- Leather scraps
- Power drill with 1/16-inch and 1/8-inch metal-rated drill bits
- Band saw
- Rotary tool with a selection of sanding drums
- Sandpaper, 180-grit and 280-grit
- Medium steel wool
- Utility knife
- Small hobby square
- Heavy scissors (for cutting leather)
- 12-inch ruler

- ✑ Craft glue
- ✑ Awl or small hole punch
- ✑ 2 heavy needles
- ✑ Heavy button thread
- ✑ Isopropyl alcohol
- ✑ Paper towels
- ✑ Anvil (or suitable substitute)
- ✑ Rivet setting tool
- ✑ Hammer
- ✑ "Internet shipping" labels
- ✑ Quick clamp
- ✑ Coping saw

Alternative Tools

- ✑ Rotary tool cut-off wheel
- ✑ Drill press
- ✑ Aviation snips
- ✑ 3-inch brass wire wheel
- ✑ Emery board
- ✑ Sewing machine capable of stitching leather
- ✑ Lockstitch sewing awl

- ✑ Steel pipe and 2-by-4s
- ✑ Strap cutter

Stage 1: Trace the Pattern onto the Slip Coupling

Step 1   Refer to Figure 5-1 for the pattern we'll use, or feel free to create your own pattern. You may be thinking, "Why not just cut the pieces with a nice straight cut? Wouldn't that be so much easier?" Figure 5-1 shows you the shape of the ocular; it is curved so it will contour to your face instead of making you look like a catfish with your eyes sticking out the side of your head when you wear them. It also indicates the location of all the holes needed.

See the page after Appendix C with Figure 5-1 set to scale so you can cut out and use as a pattern. Set your rotary punch on the smallest hole and punch the holes in the pattern where indicated.

Step 2   Clean any oils and dirt from the pipe fitting like the one shown next, and remove the nuts from the ends. If you see rubber gaskets inside the nuts, get rid of them.

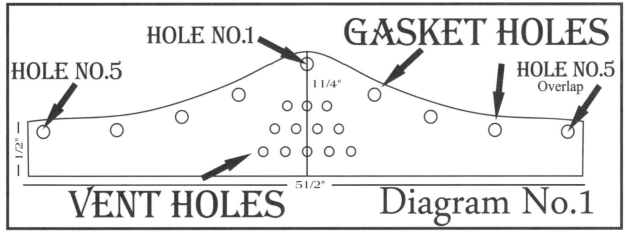

Figure 5-1   *Pattern for ocular and holes*

Step 3   Using a sharp marker, trace the pattern onto the pipe fitting. Be sure to hold the straight edge of the pattern tight against the bottom edge of the threaded part of the pipe fitting, as shown in the image on the left, and do *not* forget to mark the holes. It should look like the image on the right when you are done.

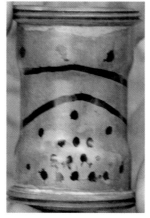

Hint
After marking the pattern and holes, let the ink dry for a few minutes to prevent your smudging it all over the piece.

Stage 2: Drill the Vent and Rivet Holes

I prefer to drill all the holes before cutting the pipe fitting into shape. It is easier to manipulate the larger piece during this stage, and since the fitting has screw threads at both ends, which keep the piece level while you're drilling. You'll need to drill two types of holes on these ocular pieces: vent holes and rivet holes. Each of the little dots you made when tracing the pattern are going to be one type of hole or the other.

The vent holes are clustered in a pattern on the widest side of the ocular. Each hole will be 1/16-inch in diameter. Vent holes help prevent the goggles from fogging up while you are wearing them.

The rivet holes are the eight around the outer edge, away from the threaded part. They should be 1/8-inch in diameter. These will hold the rivets that hold on the gasket and keep the goggles together. Yes, I know the pattern has nine rivet holes. The end holes overlap and represent one hole on the ocular.

Before you begin drilling the holes, place the pipe fitting on a hard, firm surface, preferably one you do not mind messing up. To help prevent the drill bit from "walking" while you try to drill on a curved surface, you will be punching a tiny dent at every dot. This is where the center punch comes into play.

Caution
Wear a thick pair of leather gloves when using the center punch.

Step 1   Hold the ocular with one hand to prevent it from rolling. Hold the center punch straight up and down, and press the point firmly on one of the dots. When you push all the way down, the center punch will make a loud click and a tiny dent in the wall of the pipe fitting. Repeat this step for all the dots on the pipe fitting.

Caution

You should already be wearing your safety glasses. If not, put them on now, and keep them on.

Step 2   Now it's time to drill the vent holes. (I drill the vent holes first for absolutely no reason at all.) You will need a bench vise, a power drill, and a 1/16-inch drill bit for metals (you will likely break at least one of these, so make sure you have spares) as well as a bit of leather or heavy cloth.

Put the leather or heavy cloth on the jaws of the vise to keep them from tearing up or marring the pipe fitting. Place the ocular in the vise with the hole you plan on drilling facing up.

Hint

Be careful not to over-tighten the vise on the pipe fitting; it will deform the shape of the pipe fitting. Make it just tight enough to hold the fitting firmly.

Step 3   Put the 1/16-inch bit in your drill. Place the tip of the bit into one of the little dents you made for the vent holes. Be patient as you drill these holes with a slow and steady pressure. Keep the drill perpendicular to the pipe fitting, as shown next. You will have to rotate the piece often to access the other vent holes, always keeping the drill perpendicular to the surface as you work.

Step 4   When you are done drilling all the vent holes, switch to the 1/8-inch drill bit. Now drill the rivet holes. Again, rotate the piece often and keep the drill perpendicular to the surface. This is important. When all those holes are drilled, it should look like this:

Alternative Tool: Drill Press   Of course, if you have a drill press, drilling these holes will be a lot easier. I bought a no-name press from a discount tool place for $49 on sale—in fact, it was so cheap, I bought two. You can use a drill press to do all kinds of stuff it was never intended to do. You can use it to hold a buffing wheel, as a press to compress fittings into place, and even as a mini-lathe. (I might tell you about some of that later in some of the other projects.)

If you have a drill press, you can drill the holes by following these steps:

1. Put a 1/16-inch metal-rated bit in the chuck of your drill press.

2. Place the pipe fitting on the table of the drill press—the flat part of the drill press that can be adjusted up and down. Adjust the height of the work so the bit will penetrate about 1/4-inch through the wall of the pipe fitting. This saves wear on the bit and makes it less likely to break.

3. Hold the pipe fitting firmly in place and drill the vent holes. Apply a slow, steady pressure and be patient. Mad Science takes time.

4. When you have completed the vent holes, change the bit to a 1/8-inch metal-rated bit and drill the rivet holes.

Caution

When you're holding the pipe fitting on the table, do not put your fingers inside the pipe. That would be high on the big list of bad ideas.

Stage 3: Cut and Shape the Oculars

You are now ready to cut out the oculars from the pipe fitting. The best way to do this is with a band saw. Fortunately, you can get one that will do the job from a discount tool shop for not much money (mine cost about $70).

Alternative Tool: Rotary Tool with Cut-Off Wheel   You could use a rotary tool with a cut-off wheel to cut the oculars— well, several cut-off wheels, because you *will* break them and watch the pieces fly all over the place. Make a series of connected straight cuts on the fitting, and eventually you will cut it into two pieces. That said, I believe using a cut-off wheel for this is not worth the trauma. Trust me: spend a few dollars and buy the band saw. You will be happy you did later, when you use it for other projects.

Caution
Keep your fingers out of the pipe while cutting.

Step 1   Adjust the height of the guard on the band saw to clear the top of the pipe fitting.

Step 2   *Wearing your safety glasses*, use a light metal–cutting blade to cut the pipe fitting into two halves, between the edges of the pattern tracings. Don't worry about following the lines at this point; you just want to separate the two pieces, as shown here. From now on, I'll refer to these pieces as the "oculars," not the "pipe fitting."

Step 3   Now you are ready to make the finish cuts along the edges of the two oculars. If you have a band saw and are comfortable using it up close and personal, cut along the pattern lines and remove the excess material. Be careful—while the blade might be following the line on top, the bottom of the pipe may be being cut incorrectly.

After you've trimmed the oculars, each one should look like this:

Alternative Tool: Aviation Snips
Most hardware and tool stores carry aviation snips. These snips offer a great deal of leverage and should cut the 22-gauge brass of the fitting walls with no problem. Make the cuts just barely inside of the pattern lines. Slow and steady.

Step 4   Whether you made the cuts with a band saw or aviation snips, the edges of your oculars will be pretty rough. You will need to grind and deburr the edge of each ocular. While you are at it, you will need to clean the flash—the bits of metal left from drilling and cutting—from the inside the oculars.

Use a rotary tool to hit the cut edge with a coarse grit, 1/2-inch sanding drum. Remove all the rough and sharp bits from the ocular. Use the sanding drum to clean up the cut line and make it smooth.

Step 5   Change to a medium grit 1/2-inch drum and sand the edge again.

Step 6   Sand the inside of the vent and rivet holes with the medium grit drum, making sure they are nice and smooth (after all, these are intended to go onto your face).

Step 7 Using 180-grit sandpaper, rough the inside of both oculars.

Step 8 Now use some 280-grit to clean up the outside of the holes.

Step 9 Using medium steel wool, polish the outside of each ocular. This will make the pieces all bright and shiny.

Alternative Tool: 3-inch Brass Wire Wheel Remember that I said you could use your drill press for other things besides drilling holes? You might try putting a 3-inch brass wire wheel in your drill press and using that to buff up the surface of your oculars.

Note
Do this only if you feel really confident with this method.

Step 10 Clean out any debris with a damp cloth. You do not want any of the brass bits remaining in your oculars, because these could get in your eyes later when you wear them.

Alternative Tool: Emery Board If you do not have a rotary tool, you can use an emery board to clean up your cuts and even up the edges. Then use sandpaper to clean up the holes and such. This is a lot more work, however.

Stage 4: Paint the Insides of Your Oculars

You will paint the inside of the oculars black to keep glare from the shiny brass on the inside of the goggles from obscuring your vision when you wear them.

Step 1 Wrap some masking tape around the outside of the oculars, including the threaded areas. Make sure you can access the inside with the spray paint. Trim off the extra tape with a utility knife so it looks like this:

Step 2 Use flat black spray paint to cover the entire inner surface of the oculars.

While the oculars are drying, it is time to work on the gaskets. Let the paint dry completely before you handle the pieces again.

Caution

I'm sure I don't need to tell you this, but I'm going to anyway. Apply the spray paint only in a well-ventilated place and nowhere near an open flame.

Stage 5: Cut and Stitch the Gaskets for Your Oculars

The gaskets are needed to protect your face from the metal edge of the goggle eyepieces. Select a piece of jacket-weight leather in a color you think will look really cool while you wear them around the lab ordering minions to throw the switch.

Note

The leather has a good side (the side everyone sees) and a flesh side (the side you don't usually want everyone to see). This will be important to know in this project and in general.

Step 1 Use your sharp marker and your small hobby square to mark the flesh side of a piece of jacket-weight leather into two strips, 2-by-8 inches each, as shown at the bottom of the page.

Step 2 Use heavy scissors to cut out the two strips of leather.

Step 3 Cut two, 8-inch sections of 1/4-inch diameter cord.

Step 4 Place a thin line of glue down the center of each piece of leather on the flesh side.

Step 5 Place the 8-inch sections of cord on each of the glue lines. Let this set long enough to keep the cord from rolling about—10–15 minutes.

Step 6 Fold the leather in half lengthwise so the cord is encased within.

Step 7   Now you'll stitch through the folded leather along the edge of the cord. Using an awl or a very small hole-punch, make a series of tiny holes just along the edge of the cord, each about 1/8-inch apart.

Step 8   Using a piece of heavy button thread, approximately 30 inches long, thread a needle. Run the needle through the first hole, and pull the thread halfway through. Thread a needle with the other end of the thread. Stitch the leather by passing both needles through each hole from opposite sides, as shown in the following image. This very strong stitch leaves a nice, uninterrupted line of stitching on the leather.

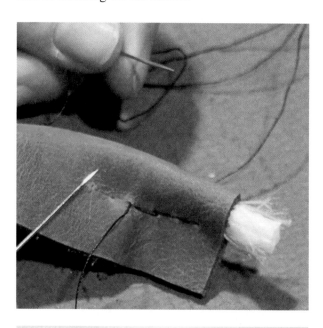

Alternative Tools: Sewing Machine, Lockstitch Sewing Awl   If you have a sewing machine that can handle leather, you can sew the leather together using a zipper foot to allow the stitch to get close to the cord. Run the stitch the length of the leather.

A lockstitch sewing awl can also come in handy for sewing leather. You can find them at leather craft stores, fabric stores, and most hardware stores. It is pretty simple to use and comes with instructions.

Whatever method you use, the finished pieces should look like this:

Stage 6: Attach the Gaskets to the Oculars

The paint inside your oculars should be dry enough to handle. I'll show you how to attach the gaskets to the oculars. You might be thinking, "How hard could it be?" but a few subtle things worth noting will make the goggles look more professional and fit better. The gaskets should be a bit too long to go all the way around the oculars' edges, and you will, of course, be cutting off some of the excess.

Step 1   Place one of the leather gasket strips on the table in front of you, lengthwise, with the thicker edge containing the cord away from you.

Step 2   Measure about 3/4 inch in from the right side and about 1/8 inch from the bottom edge, and use your sharp marker to make a dot.

Step 3   Rotate the punches on your rotary punch to the second smallest setting, and punch a hole through both flaps where you just made that dot. It should look like the following image.

Step 4 Place the ocular on a table with the screw thread part down. Make sure the hole on the tallest part of the ocular is facing you.

Step 5 Separate the flaps of the leather gasket. Then sandwich the wall of the ocular inside the flaps of the gasket. Line up the hole you made in the gasket with the top hole on the ocular: let's call that hole number 1, as shown back in

Figure 5-1 on your ocular hole pattern. From the inside, insert a small brass rivet post into the hole. Do not put a cap on it; this rivet is just to hold the gasket in place for now.

Note

This next part is difficult to explain, so please bear with me and pay attention to the pictures.

Step 6 Normally, you don't want to make marks on the good or top side of the leather, but sometimes there is no way around doing this, and this is one of those times. When last we left our ocular, it was sitting on the table with the gasket barely started around it, as shown previously. Leave it there. Guide the gasket edges to sandwich the wall of the ocular for about an inch or so around the ocular to make sure the end of the gasket you intend to mark is at the correct angle.

Step 7 Place your small hobby square on the table in front of the ocular with the line of the upward straight edge passing through hole number 1, as shown here:

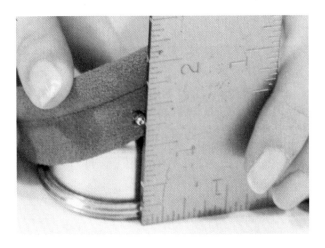

Step 8   Use your sharp marker to mark the gasket along the straight edge—but *only* on the thick part containing the cord, like this:

Step 9   To mark the flap that sandwiches the ocular, you will draw a line down from the bottom of the previous mark and toward the outside tip, leaving plenty of room for the hole. Just make sure it looks something like this:

Step 10   Remove the gasket from the ocular, and cut it along the lines you just drew so the end looks like this:

Step 11   Use a marker to color the white end of the cord to keep it from showing too much.

Step 12   Place the ocular on the table, and use your marker to draw lines straight down from each hole to the screw threads, like so:

Step 13   Slide the gasket back onto the ocular, as it was in step 3. Again, add a rivet post in hole number 1 to hold the gasket in place, but do not put a cap on it. Sandwich the ocular wall between the flaps and push the gasket down so its edges cover the next hole, overlapping it by a tiny bit more than 1/8 inch. The line you drew on the ocular indicates the location of the next hole. Mark the spot where the hole is located by putting a dot about 1/8 inch from the gasket's edge, just over the line on the ocular, like this:

Hint

When you punch holes through the leather gasket, make sure the bottom edges are flush and that you are going through both pieces.

Step 14   Slide that section of the gasket off while leaving the post in the hole number 1. Pivot the gasket for better access, and punch a hole at the mark you just made with the smallest punch on your rotary punch. Slide the gasket back down to line up the gasket hole with the ocular hole. Place a post in this hole, without a cap.

Step 16   Mark and punch the last hole, but don't put a stud in it.

Note

You are going to cut off the extra bit, but don't just start chopping away just yet.

Step 15   You can probably guess what will happen next. Yes, you will do the same for the next hole. It should look like the following image. Continue doing this all the way around, *except for the last hole*—not the overlap for hole number 1, but the one before that: the penultimate hole, if you will.

Step 17   Remember when you put the gasket at hole number 1, and marked where to cut the end? Well, you are going to do the same thing on this end. Take the stud out of hole number 1 and the one next to it, and gently peel off the gasket, as shown at the top of the next page.

Step 18   Place the uncut end of the gasket around the ocular, sandwiching the wall between the flaps just like before. Follow the line that indicates the position of hole number 1, and at 1/8 inch from the edge—this is all very familiar, I'm sure—use your marker to mark the hole location in the uncut edge of the gasket.

Step 19   Use the rotary punch to punch the last hole in the gasket, and replace the rivet post in the hole. You're back at hole number 1.

Step 20   Place your small hobby square on the table in front of the ocular, with the line of the upward straight edge passing over hole number 1. Mark only the part of the loose end of the gasket containing the cord—yes, just like before. Mark

the flap the same way you did on the other end, to make sure you have covered hole number 1, plus a tiny bit to grab onto. Make sure it all looks like the image just shown before you cut it off. Take a look at the illustration with step 22 to get another look at this piece. When you're satisfied, cut it off.

Step 21   Don't forget to use the marker to color the white part of the cord.

Step 22   Now sandwich the ocular wall at hole number 1 with *both* ends of the gasket, and put the post through both ends of the gasket. It should like this:

Step 23   Now take out all the studs and remove the gasket completely.

Step 24   Use some isopropyl alcohol and a paper towel to remove the marker lines from the ocular. Wrap the gasket around the ocular.

Caution

After you've cut and punched both gaskets, be careful to keep track of which gasket goes on which ocular. And be sure you don't cut both gaskets and its holes using the same ocular; the holes won't match up on one of the oculars if you do this.

Step 25   Repeat steps 1–24 for the other ocular and gasket.

Step 26   Now you are ready to rivet the gaskets to the oculars. Looking down from above, with the threaded end of the ocular on the table, working clockwise, you should count eight holes all the way around. Make sure a small double cap rivet stud is set in every hole—*except* for hole numbers 1 and 5 (see Figure 5-1; I'm referring to *both* hole number 5s)—and place a cap on each. (If there is a post in hole 1 and/or 5, remove it.)

Step 27   Time to set your rivets. This can be a bit tricky, because there is nothing supporting the back of the rivet. Fret not, however, because there are clever ways to get around this little setback. If you happen to have an anvil, preferably one with a small beak (the pointy bit of the anvil), place the ocular over the beak with the rivet you intend to assault facing straight up.

Step 28   Pick up your rivet setting tool. Notice that one end is slightly concave. Place that end on top the first rivet cap. Pick up your hammer and give the top of the tool a good solid thwack, like you can see me doing in the illustration. You don't need to try and take the beak of the anvil off, but do let the rivet know just who the mad scientist is around here. Repeat this for all five remaining rivets (no rivets in hole numbers 1 and 5, remember).

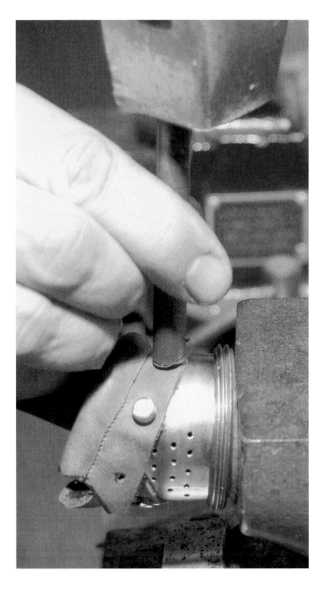

Note

You can pick up a small anvil at a hardware or tool store pretty cheap. You don't need a big one—maybe only 10 or 20 pounds. An anvil will provide a great advantage when you're working on other projects. If the idea of an anvil that size is a bit daunting, try a "mini" anvil that weighs about a pound; you can find them online. They are a great deal less expensive to ship than the heavyweights. You will need to mount the anvil to a table—or perhaps a tree stump if that's your work surface of choice.

Alternative Tools: Steel Pipe and 2-by-4s   If you do not have an anvil, or if you really did break the beak off your anvil murdering your first few rivets, you can use a piece of heavy steel pipe, at least 8 inches long and at least 1 inch in diameter. Find two 2-by-4s and place them flat on the table about 4 inches apart. Place the pipe section like a bridge between the two pieces of wood. Add a screw into each block of wood on both sides of the pipe to keep the pipe from rolling around. Remove one end of the pipe from the wood block, and slide the ocular over the end and all the way to the middle. Place the pipe back on the wood blocks and between the screws. Then hammer the rivets in the same fashion you would if you had an anvil.

Step 29   Repeat steps 26 and 28 for the second ocular.

Stage 7: Attach the Nose Bridge

The nose bridge is the leather piece that goes between the oculars and over your nose. The length of this piece dictates the distance between the oculars. If you have a wide face, make it a little longer than stated here; if you have a quite narrow face, make it a little shorter. Read through the instructions, and hold it all up to your face before you cut. This stage is almost too easy to explain, but I will anyway.

Step 1   Cut a piece of the belt-grade leather into strips 1/2 inch wide and about 2 feet long. Use a ruler or some other long straight edge to mark out your pieces on the back side of the leather, and then use a pair of leather cutting scissors to cut the strap—if you do not have leather scissors, use the heaviest scissor you can get your hands on.

Hint

I used an old belt from a thrift store for the goggles featured in the book. It happened to be black, so you cannot easily see a black marker on it. The good news is you can now buy silver markers that mark on black with ease. I highly recommend them.

Alternative Tool: Strap Cutter   If you have a strap cutter, your world will be a much happier place. You can get one in any leather craft shop or online. A couple different types of strap cutters are available, and they usually come with instructions written by trained professionals. Please read those instructions thoroughly and avoid the trauma of the trial and error, and error, and first aid method of learning. This tool will make strap making an absolute breeze. Many of the projects in this book have straps of some sort, so your purchase will be money well spent.

Step 2   Cut a piece from the strap about 2 inches long.

Step 3   Use your rotary punch to place a hole 1/8 inch from each end of the nose bridge so it looks something like this:

Hint

Whenever you cut a belt or strap, try cutting the corners down just a bit. It will make the piece look just a bit more...intentional. I also might cut a slight arch into one of the long sides of the nose bridge. This gives the goggles a definite top and bottom. The arch will fit directly over the nose.

Step 4   Line up one of the holes on the nose bridge with hole number 5 (remember when you numbered the holes while fitting the gasket?) on the ocular. Make sure the "flesh side" of the leather nose bridge faces up and that the strap is rotated so that it sticks out toward the front of the ocular, past where the lens will be (see the next image).

Step 5   Place a medium-size double-cap rivet post though hole number 5 from the inside of the ocular. Pass it through the hole in the gasket and the hole on the end of the nose bridge. Put the cap for a *small* rivet on the post, as shown here:

Step 6   Place the ocular on the beak of your anvil, or on the pipe, just as you did in the last phase. Place the rivet setting tool on the cap and whack it with the hammer.

Step 7   Place the hole on the other end of the nose bridge over hole number 5 on the other ocular. Make sure it is in the same position as the hole on the first ocular. The nose bridge should arch away from the face when you wear the goggles.

Step 8   Now repeat the bit where you put the rivet through the holes and all that. Whack the rivet head. See how they are starting to look all goggle like? They should look like this:

Stage 8: Attach the Head Strap

This is where we make the strap that holds the goggles onto your head. I will show you how to add a buckle to a strap and to make the tongue (that's the part with all the holes for adjusting the fit). You might need these skills for other projects—these are good things to know anyway.

Note

Make sure you are measuring away from the ends of the pieces when making all these holes.

Step 1   Cut two pieces from your 1/2-inch leather strips. Cut one to 12 inches, and write a small "B" on the flesh side. Cut the other one to 15 inches, and label it "T" on the flesh side. Always remember to mark your leather on the flesh side whenever possible.

Step 2   Then cut two more pieces at 2 inches each. Remember what I said about cutting the corners, too.

Step 3   Using your rotary punch, make a rivet hole (to fit a large rivet) 1/8 inch from each end of each of your 2-inch pieces. These are your "ring straps."

Step 4   Loop one of the ring straps through one of the "D" rings, with the flat bit touching the flesh side of the strap, so it all looks like this:

Step 5   Starting from the inside of one of the brass oculars, add a post of a large double-cap rivet in this order:

- ❧ one of the end holes on the ring strap
- ❧ the inside flap of the gasket
- ❧ hole number 1 of the ocular
- ❧ the outside flap of the gasket
- ❧ the top side of the other hole of the ring strap

The ends of the ring strap should sandwich the wall of the ocular inside hole number 1 and around to the outside of the gasket. Make sure the ring strap is not twisted, and that the top side of the leather faces out. Place a large rivet cap on the top of the post; it should look like this:

Step 6   Place the ocular on the anvil or pipe. Place your rivet setter on the cap, grab your hammer, and give it a good whack.

Step 7   Follow steps 4–6 for the other ring strap and ocular. I hope you like doing this because you will be doing it a lot.

Hint

I'm going to interrupt things a bit here. The next few steps walk you through how to put a 1/2-inch buckle on the end of a 1/2-inch leather strap. This is pretty useful information, as I tend to use straps on a lot of my projects.

Step 8   Grab the 12-inch strap labeled "B." Place a mark for a hole at 1/8 inch from one end— this will be the buckle end. Measure 3/4 inch from that mark and make another mark. Measure 1/2 inch from that mark and make another mark. Then measure 1/2 inch from that mark and make another mark, and then, you guessed it, measure 1/2 inch from that last one and make a mark. Now make one more mark 3/4 inch from the very last mark. You should have six marks on this end of the strap.

Hint

See the page after Appendix C with Figure 5-2 set to scale so you can cut out and use as a pattern, rather than making the measurements yourself.

Step 9   Set your rotary punch to the second smallest sized hole, and punch holes at the six marks you just made.

Step 10   This is another tricky part. See those two holes in the middle of the six you just made? You are going to connect them. Grab (carefully) a utility knife with a brand new, sharp blade and your ruler or a rigid straight edge. Place the strap flat on the table, running left to right.

What Buckle?

Take a good look at your buckle. The type of buckle I recommend for this project has three distinct parts. I probably have the names wrong, but it's my book and I will call them what I want. So there.

- The *prong* is the part that sticks through the holes on a belt.
- The *bar* is the bit on which the prong swivels.
- The *frame* is the thing that goes around the outside. On many buckles, one section of the frame is a bit thicker than the other. We will call this the *front* of the frame.

So now that we have all that out of the way, we can get on with putting the buckle on the end of the B strap.

Place the straight edge against the bottom of the two middle holes, but so you can see the holes. Very slowly and carefully, cut along the straight edge, connecting the two holes. Rotate the strap 180 degrees and place the straight edge along the bottom of the holes again, and cut across like before. You have now turned the two holes into a slot. This end of the strap should look like Figure 5-2 when finished.

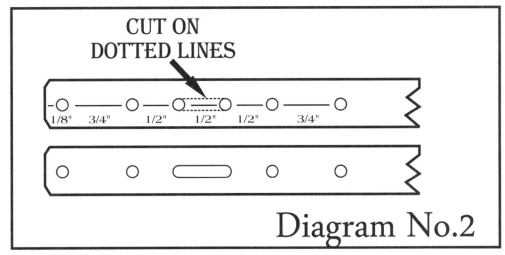

Figure 5-2   *Pattern for 1/2-inch strap buckle placement*

Step 11   Holding the buckle horizontal, put the buckle end of the B strap through the back end of the frame of the buckle. Stick the prong of the buckle straight up and into the slot you cut into the strap. Put the end of the strap down through the front part of the frame so it wraps around the bar.

Step 12   Now there is, of course, more riveting. Place the post of a medium double cap rivet through the good side of the hole just to the front end of the slot. Wrap the strap around the bar and put the tip of the stud through the back side of the hole on the other side of the slot. Put the riveted leather over the flat part on your anvil or the top of a vise. Add a small cap and whack it.

Step 13   From the other end, slide a 1/2-inch "D" ring up the length of the strap to the rivet just behind the buckle. Make sure the round part is up and the flat part is sandwiched between both pieces of the leather strap. Pass another medium-size rivet stud up through the hole in both pieces of leather behind the "D" ring, trapping it between the two rivets. Time to cap and whack. Again use a small cap. Now you have a buckle on the end of a strap. It should look like the next image. Please make sure it does.

Step 14   Use your sharp marker to mark the location of two more holes in the buckle strap—yes, I am going to tell you where they go. Mark one hole 1/8 inch from the other end of the strap. Measure 1 1/2 inches from the center of the end hole and make another mark. Punch rivet holes at these two marks. This is now called the "D" ring end of this strap.

Step 15   From the outside, thread the "D" ring end of the B strap through the "D" ring on one of the oculars, wrapping the B strap around the ocular "D" ring. Line up the two end holes on the B strap, add a medium rivet stud through the good side and up through the back side of the other hole on this end, and now, you guessed it, cap it and whack it.

Step 16   On to the T strap. Make a hole 1/8 inch from the end—it does not matter which end yet. Now make another hole 1 1/2 inches from the center of the last hole. This is now the "D" ring end of this strap.

Step 17   Measure 3 inches from the hole you just made and make a mark. Measure 3/4 inch from there and make another mark. Keep doing this—measure 3/4 inch and make another mark—all the way to the end of the strap.

Step 18   Set your punch to the second to largest hole size. Punch a hole at each of the marks you just made. These holes are larger than the rivet holes; the prong of the buckle will go through them. Cut the end just after the last hole to a point to make it easier to insert into the buckle.

Step 19   Now you are going to attach the "D" ring end of *this* B strap onto the ocular. It goes on the exact same way as the other one, but I'm going to tell you how to do it again just in case you have forgotten. Thread the "D" ring end of the T strap through the "D" ring on the ocular from the outside, wrapping around the ocular "D" ring. Line up the two end holes, add a medium rivet stud through the good side of the end hole, up through the back side of the other hole on this end, and cap it and whack it.

The straps are done! Are you getting excited now that your goggles are near done and looking like the image shown below?

Stage 9: Add the Lenses (Finally)

You need your piece of acrylic now, which you will cut into two discs, 1 3/4 inch in diameter. If you have an industrial cutting laser, you probably do not need me to tell you how to cut these discs. So I will discuss a more low-tech method.

Hint

The acrylic typically comes with a protective plastic or paper coating. Resist the urge to peel this off. It will keep the lenses from getting all scratched up and make it easier to mark on.

Step 1   The easiest way to get this just right is to acquire "Internet shipping" labels for your printer. Use whatever art program (or a compass)

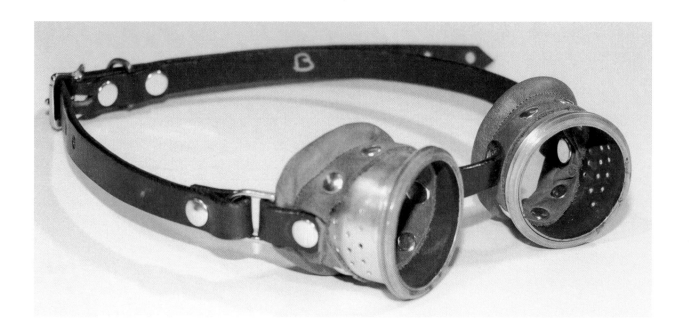

you have to draw two, 1 3/4-inch diameter circles. Make the line width small and print them onto the labels. Peel off the label backing and stick them to the acrylic sheet—both circles on the same side and near the edge would be preferable. Trim off the excess label so it looks like this:

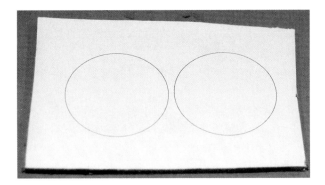

Step 2   Clamp the acrylic piece on to the edge of a table with, at most, half of a lens hanging off the edge of the table. Use a coping saw to cut out the lenses by following the edge of the circles. Move and rotate the piece as needed, but keep about half the circle on the table, as shown here:

Step 3   Clean up the edge with some sand paper or a belt grinder.

Step 4   Peel the protective coating off the lenses.

Step 5   That's it. You should now have a lovely pair of lenses. Here's a pair made from translucent red acrylic:

Stage 10: Final Assembly

Step 1   Place the lenses inside the nut ends of the ocular pieces (that you removed at the beginning of the project).

Step 2   Screw the end nuts, the rims, onto the oculars.

Step 3   It might seem a bit anticlimactic, but that's it! Your goggles are done.

Note

A variation of this goggles project—a monocle—is available for download at www.mhprofessional.com/steampunk.

Chapter 6

Calibrated
Indicator Gauges

We made the controls of the Ornithoptic Boarding Conveyance as simple as possible. As not to confuse our Marines with a great many numbers, which they would not understand anyway, the pressure gauge indicator has only three values: Falling (pressure too low, please adjust), Flying (pressure just right), and Dead (pressure too high but no need for further adjustment). We recommend the Marines try and keep the pressure at "Flying" whenever possible.

—From transcript of Col. Waiselthorpe testimony,
during the Air-Admiralty's inquest of the "O.B.C Affair"
and review of the *Areo-Naval Boarding
Actions Handbook* by Col. Waiselthorpe

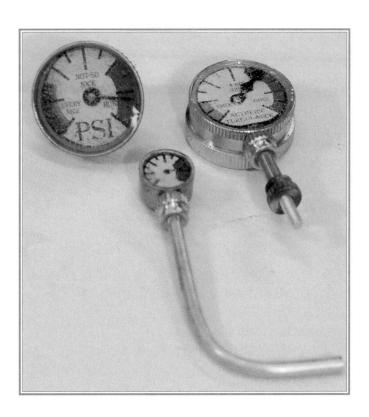

*T*hrough constant and continuous scouring of local flea markets, antique shops, and hardware stores, I have over the course of weeks, months, and years amassed a large number of both antique and modern gauges. These range from steam pressure gauges, to temperature gauges, to voltmeters, altimeters, speedometers, and beyond. And although these antiques are marvelous in both form and function, they are very often too large for small portable projects. An 8-inch diameter steam pressure gauge salvaged from a derelict locomotive is not generally conducive to adaptation as a portable, wearable accessory.

To that end, I have devised a method for the construction of gauges that can be tailored to size and scale for relevant projects. The variety and versatility of this type of gauge design is nearly limitless. Although the gauges we build in this chapter are no larger than 1½ inch, the same process can be used to build gauges of any size or scale up to and including the size of a coffee table—hmm...come to think of it, that is not a bad idea. I might just have to make one of those someday.

The reader would be well advised to read through the entirety of this chapter before undertaking this project, because I have included several alternative methods for constructing and completing these gauges. If you do not read through to the end, you might end up taking your project down the wrong path and find yourself at a dead end, where another path would have been far simpler, as determined by what tools and materials you have on hand.

Project Description

This chapter will take you through the process of creating artificial gauges in several sizes. Although these gauges are not intended to be projects unto themselves, they are incredibly versatile and can be used to add detail, character, and ornamentation to most of the other projects in this book.

A gauge can also impart a sense of tension and danger to any artifact. How frequently have you read or watched some story wherein the hero or heroine is in a race against time to complete some task before the gauge indicates some infernal device has reached a critical stage? Gauges convey the sense that the fantastical mechanisms and devilish devices we construct as part of the Steampunk aesthetic are actually functioning (or malfunctioning) on their own.

You could, of course, simply go out and purchase a gauge from a hardware center and then simply repaint the housing. But this chapter will show you how to make any gauge of any size that says whatever you want. Let's face it—if you are content with a gauge reading "PSI" or "Oil Pressure," they are not difficult to find. But unless something has drastically changed, I would consider it unlikely that one could pick up gauges at a big box store that calibrate aetheric turbulence, temporal distortion, earth atmospheric integrity, or even death ray charge.

What You'll Need

To complete the project as described in this chapter, you will need to get your hands on the following materials and tools.

Materials

- Metal caps, lids, or cups (one per gauge)
- Two-part, 5-minute epoxy
- 1, 6-32 × 3/4-inch brass machine screw
- 2, 6-32 brass nuts
- White glue
- 4 inches of 3/16-inch aluminum tubing

∽ 1, #6 brass washer

∽ 1, 6-32 × 1 1/2-inch brass machine screw

∽ 2, 6-32 knurled nuts

∽ 4 inches of 3/16-inch brass tubing

∽ Modeling clay

∽ Glossy paper or sticker stock

∽ Small, double-capped rivet (optional)

∽ 3 small indicator hands (store-bought or handmade)

∽ Spray adhesive

∽ Cardboard

∽ Two-part clear acrylic epoxy or acrylic sheet or Plexiglas, at least 1/16 inch thick

Tools

∽ Safety glasses

∽ Awl or spring-loaded center punch

∽ Power drill

∽ 9/64-inch drill bit

∽ Fine-point permanent markers

∽ Bench vise

∽ Small tubing bender

∽ Band saw or hacksaw or rotary tool with cut-off wheel

∽ Caliper (dial, digital, traditional, or substitute)

∽ Scissors

∽ Ruler

∽ Translucent paints, markers, or highlighters (in red, yellow, and green)

∽ Craft knife

Alternative Tools

∽ Computer and design software

∽ Drafting compass or circle templates

Stage 1: Construct the Housing and Mount

First of all, you need to determine how you intend to use the gauge. Do you need an aether-variance calibrator intended to be mounted onto the side of your goggles? Or perhaps you need to monitor air-pressure fluctuations on the side of your high-altitude mask. Or will the gauge track the variations in temporal interference and chrono-static on your custom-made Tesla-pod? Whatever the purpose of the gauge, your decisions at this stage as to where it will be attached will determine its size, design, and construction.

Although all of that might sound complex, what it all boils down to at this stage is the gauge's housing. As you can see in the illustration, I have collected a number of suitable candidates.

These housings can be almost any cup-shaped metal object—small sliding door pulls, copper end caps for plumbing, threaded pipe caps, or little brass cap-like pieces found in a box of old junk (not that anything in a Mad Scientist's Steampunk laboratory should ever be considered junk). I also have a metal cap (which I think came from a bottle of hot sauce)—any metal cap would do, such as the cap from a ketchup bottle, steak sauce bottle, or whatever. In this book, we will be making a few different variations on this design, but the same principles would apply no matter which style cap you have chosen to use as your gauge housing.

To illustrate the variety and versatility of this design, I will show you how to make a few different gauges with three distinct mounts: a surface mount made with a hot sauce bottle cap, a lollypop mount made with a brass garden hose cap, and a 90-degree mount made with a 3/8-inch copper tubing cap. Read through the steps for each type of gauge, and then decide which of these mounting designs is best suited to your project and follow the directions to create that style.

Style A: Surface Mount

Step A-1   Don your safety glasses. Locate the center of your cap (in this case our hot sauce bottle cap) and dimple the inside of the cap using an awl or spring-loaded center punch. Then drill out the hole from the inside with a power drill fitted with a 9/64-inch drill bit.

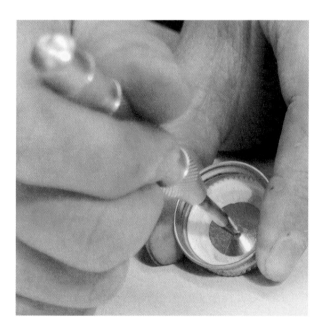

Note

We are using a 9/64-inch drill bit because it is compatible with the 6-32 machine screws we will be using for the mount, but your scale might, of course, be different. Let us assume, however, for the remainder of the chapter, that you will be making gauges that will be identical to the ones in this book in every regard.

Step A-2   Slide a 6-32 × 3/4-inch brass machine screw through the hole from the inside of the cap, and secure it in place using a 6-32 brass nut. Use a dab of adhesive inside the nut to minimize the potential of the nut vibrating loose through use and wear.

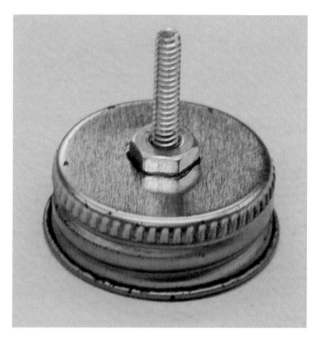

Step A-3   We are now going to disguise the machine screw thread by hiding it within a short 1/4-inch-long piece of 3/16-inch aluminum tubing (available from most hobby shops). I like this design because it makes the gauge more fully three-dimensional so that it does not appear to be simply glued on to your project.

Note

Step A-3 is necessary only if you want your gauge to stand out and away from whatever you are attaching it to. If, on the other hand, you intend for your surface mount to be flush to the surface, you can skip this step and the next.

Step A-4   Secure the tube in place using a #6 brass washer and another 6-32 brass nut. Secure the nut with adhesive.

Style B: Lollipop Mount

For the lollipop mount, I have chosen to use a brass garden hose cap (available from most national hardware chain stores or garden centers).

Step B-1   Using a sharp marker, mark where you want to attach your mount on the outer edge of the housing—roughly the midpoint of the depth of the cap. Once you have marked the position, dimple the location using the spring-loaded center punch or awl.

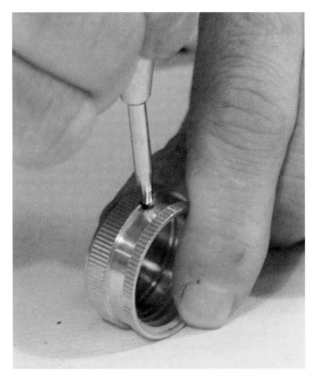

Step B-2   Don your safety glasses. Clamp the housing tightly in a bench vise and drill out the hole using a drill fitted with a 9/64-inch bit.

Note

Just like before, we are using a 9/64-inch drill bit because it is compatible with the 6-32 machine screws we will be using for the mount, but your scale might, of course, be different. Let us assume, however, for the remainder of the chapter, that you will be making gauges that will be identical to the ones in this book in every regard.

Step B-3   Slide a 6-32 × 1 1/2-inch brass machine screw through the hole, and secure it in place using a #6 brass washer and a 6-32 brass knurled nut. Be certain to use a dab of adhesive inside the nut to minimize the potential of the nut vibrating loose through use and wear.

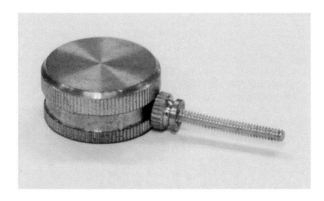

Step B-4   We are now going to disguise the machine screw thread by hiding it within a short 1/2-inch-long piece of 3/16-inch brass tubing (available from most hobby shops).

Note

Step B-4 is necessary only if you don't want your gauge to look rather daft just stuck out to the side like that and wish it to stand out and away from whatever you are attaching it to. If, on the other hand, you just want your gauge to be mounted flush to the surface, you can skip this step and the next.

Step B-5   Secure the tube in place using another 6-32 brass knurled nut. Secure the nut with adhesive or white glue.

Style C: Ninety-degree Mount

For the 90-degree mount, I have selected a 3/8-inch copper tubing plumbing cap, available from most places that sell plumbing supplies.

Step C-1   Using a sharp marker, mark where you want to attach your mount on the outer edge of the housing. This should be at roughly the midpoint of the depth of the cap. Once you have marked the position, dimple the location using a spring-loaded center punch or awl.

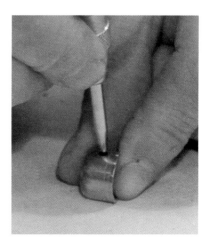

Step C-2   Don your safety glasses. Clamp the housing tightly in a bench vise and drill out the hole using a drill fitted with a 9/64-inch bit.

Note

Yet again, we are using a 9/64-inch drill bit because it is compatible with the 6-32 machine screws we will be using for the mount, but your scale might, of course, be different. Let us assume, however, for the remainder of the chapter, that you will be making gauges that will be identical to the ones in this book in every regard.

Further Note

Unless you intend to use screws that are shorter than the internal width of your cap, you might want to drill straight through both sides of your housing so that a machine screw can be slotted straight through.

Step C-3   Slot a 6-32 × 1 1/2-inch brass machine screw through both holes, and secure it in place using a #6 brass washer and a 6-32 brass knurled nut. Be certain to use a dab of adhesive or white glue inside the nut to minimize the potential of the nut vibrating loose through use and wear.

Step C-4   We next want to construct our 90-degree angle by taking a piece of 3/16-inch soft brass tubing (available from most hobby shops). Use a small tubing bender (available online or through good hobby shops and hardware centers) and bend your brass tubing to a 90-degree angle.

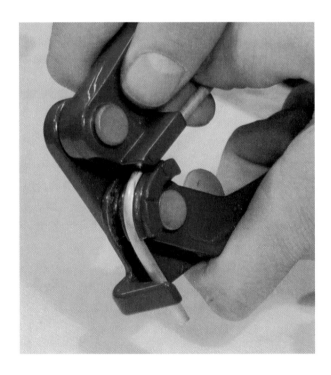

Step C-5   Place the gauge with the mount screw next to the bent tubing so that the end of the screw is equidistant with the beginning of the bend. On the tubing, mark the distance to the base of the knurled nut, as shown next.

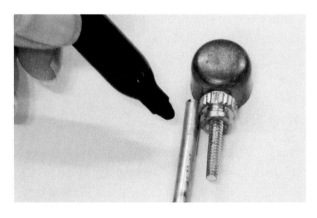

Step C-6   Cut off the tubing at the mark using a cut-off wheel on your rotary tool, or with a band saw or small hacksaw.

Step C-7   Mix up a small amount of two-part, 5-minute epoxy and coat the exposed machine screw thread before slotting the screw into the bent tubing.

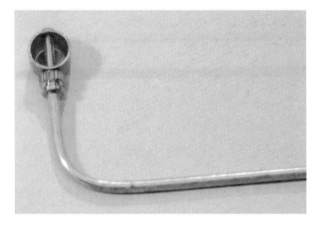

Hint
Disposable wooden craft sticks and a scrap of cardboard are ideal tools for mixing up the two-part epoxy.

Step C-8   Determine the desired length of the return on the lower part of the bend. Allow an extra inch or so of leeway (you can always cut it shorter later), and cut the tubing to length using the same tool you used in step C-6.

After you've attached the mount for your gauge, you can move on to creating the face.

Stage 2: Measure, Make, and Install the Gauge Face

Here we are going to transform our plain metal cup into something that just about anyone would recognize as a functional gauge. Bear in mind that the three examples I show you in this chapter are simply that—examples—and you should feel free to explore, experiment, and get creative with your project. You do not necessarily need to reproduce these examples, but you can follow the step-by-step process as a guide to your own creative designs.

Step 1   First and foremost, you need to measure accurately the internal diameter of your cup. You can do this in a number of ways, depending on the type of cup/cap you are using and the tools you have available.

Method A: Digital Caliper   If you have a good dial or digital caliper and know how to use it, this is the ideal tool for taking this internal measurement. The calipers shown here are typical of the type of dial caliper that I find so frequently and incredibly useful. If you don't have one of these, I highly recommend that you acquire one (try your local hardware store). First read the instructions, and then run around your house measuring all sorts of things you never before needed to measure. When you return to your cap/cup, be certain to use the internal caliper, as shown (as opposed to the external caliper).

Method B: Traditional Caliper
Alternatively, you could use a traditional caliper. These can frequently be found at any decent flea market or antique shop. It is always worth digging through old tool boxes and "junk" piles, because antique tools are often more suited to Steampunk purposes than their modern or contemporary counterparts. If it worked for the task 100 years ago and has not suffered any debilitating damage in the interim, it should likely still perform its intended purpose today. If you are unfamiliar with the use of this sort of caliper, simply use the internal calibration end to span the interior diameter of your cup/cap, and then use a ruler to measure the opposite end of the caliper, which should be exactly the same as your internal dimension, as shown:

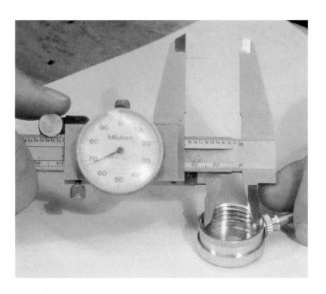

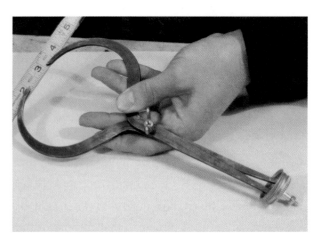

Hint

A true Steampunk genius is always able to improvise and adapt by using his or her knowledge of physics and mathematics to create the right tool for the job. This is generally referred to as "MacGyvering" a solution. (If you do not get that reference, you are too young and I feel sorry for you.) In this case, we can MacGyver a solution by using a standard pair of scissors or a drafting compass, and opening the points within the interior of the cap/cup. Fix the scissors or compass into an open position using a piece of tape, then remove them and measure the distance between the outside of the points.

Note

At the risk of sounding patronizing, do not work from the end of your ruler when taking precise measurements, because the end of the ruler is usually the least accurate part of the measuring device.

Further Note

If you have purchased a metal hose or plumbing cap for your project, the exact internal diameter of your cap/cup might well be indicated on the thing's packaging. You do still have that packaging, right?

Method C: Standard Ruler   If you do not have a decent caliper (or you don't feel like improvising one), you might have to rely on a standard ruler for this purpose. Getting an internal measurement can, of course, be a tricky prospect (at least to a high degree of accuracy), but as this is, after all, a purely ornamental piece, and as such will not be under any actual pressure or whatever, a slightly loose fit might be acceptable to you. Hold the ruler over the cup across the widest point. And "eyeball" the inner diameter measurement. It's that easy.

Method D: Transference   If you are not content with the ruler method, I suggest you use what I will call the "transference method." To do this, take some modeling clay (which will be used later in this project) and lay out a smooth, flat piece that is larger than the external diameter of your cup/cap. Press the cup/cap face into the clay, leaving a clear impression. It should be a relatively simple matter to measure the internal diameter of the impression you have made in the clay with a ruler.

Step 2   Once you have used one of the above methods (or one of your own devising) to measure accurately the internal diameter of your cap/cup, you are ready to create the gauge face. To do this, you again can use a couple of viable methods: You can use design software and a computer, or you can hand draw a design.

Method A: Computer-aided Design

I have no way of knowing what sort of graphic design software you might have installed on your computer, but I have to assume that you know how to use whatever software you installed. It is beyond the scope of this book to teach you graphic design or computer-aided drafting, so, to make this easier, I will assume that you have CorelDRAW 12 (because that is what I am using for this project). You could, of course, use Photoshop or any other design software for this.

1. Render a circle that is exactly the same size as the internal measurement for your cup/cap (as determined in Stage 2: step 1). In this case, I am creating a circle that is 1 1/16 inch (1.062 inch) in diameter.

2. I now want to render all the gauge calibration intervals (which I usually refer to as "all the little tick-y marks"). I begin by creating a vertical line (using the freehand tool) about 1/4 inch (0.25 inch) tall. I position the top

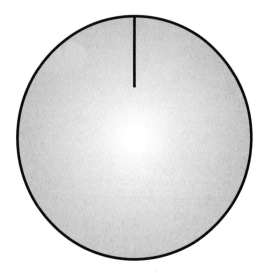

edge of my tick-y at the 12 o'clock position within my circle.

3. Then I make radial copies of my tick-y mark. If you're following along with CorelDRAW, go to the Arrange | Transformations | Rotate window, and a docker should be visible on the right side. Use the Pick tool to select the tick-y, and press F10 to enter Node Edit mode. In the center of the tick-y should be a dot with a circle around it. Click that dot and drag it to the center of the gauge face circle. In the Transformations docker, you should see a window for Angle. Set the angle to 45 degrees. Now click Apply To Duplicate three times in the docker window, and your design should look something like this:

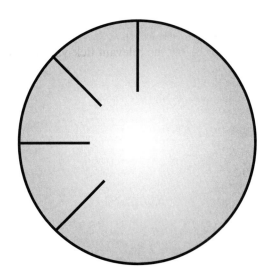

4. Select the Pick tool. Select your tick-y in the 12 o'clock position again. Return to the Transformations docker, change the angle to −45 degrees, and click Apply to duplicate three times. Your design should now look like the image at the top of the next column.

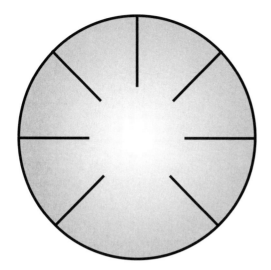

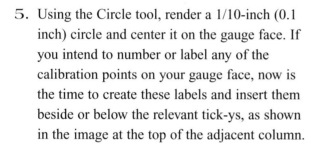

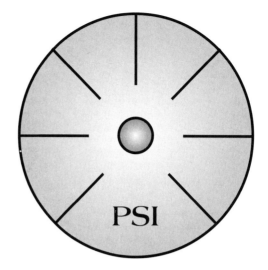

5. Using the Circle tool, render a 1/10-inch (0.1 inch) circle and center it on the gauge face. If you intend to number or label any of the calibration points on your gauge face, now is the time to create these labels and insert them beside or below the relevant tick-ys, as shown in the image at the top of the adjacent column.

   This is perfectly suitable as a simple, straightforward gauge face. And for the smaller size gauges (such as the 1/2-inch gauge in our examples), you probably do not want to complicate this any further, lest the additional marks or numberings, and so on, become illegible at first glance. If, however, you are creating a larger size gauge and you

want to ornament your gauge face further with 1/2-inch and 1/4-inch tick-ys, or with numbers and words, now is the time to render those as well. There are no hard or fast rules for this, so feel free to experiment and have fun with it.

6. Repeat this process for each gauge you intend to make. When you're creating multiple gauges of varying sizes, simply select your finished gauge face and copy/paste it, and then change the size by selecting the duplicate and altering the measurement in the object size window. Repeat until you have gauge faces in each size necessary for the gauges you have made, as shown below.

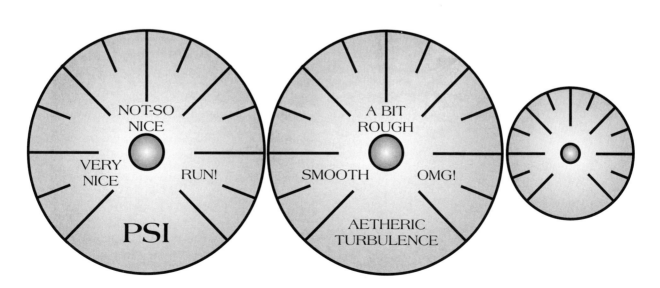

7. Save your work and print out your graphic file onto standard letter paper. This is a test print to ensure details and size. If you're satisfied that they will work correctly, print out a higher resolution version onto glossy photo-stock paper or onto sticker stock (such as that used for mailing labels).

Method B: Hand-drawn Design For that truly handmade look, you can carefully mark out your gauge face by hand using fine-point markers.

1. Using a drafting template or a compass, trace out a circle that is the same diameter as the interior of your cup/cap housing. You could, of course, simply use your cup/cap housing as the template and trace around it, but bear in mind that the gauge face needs to be the same size as the *inner* diameter of the cup/cap, so an allowance would need to be made for the thickness of the housing walls or for any lip or rim that might exist on your cup/cap.

2. Mark your gauge calibration intervals (the little tick-y marks) in light pencil, being careful to ensure even spacing. Also be certain to mark your center point.

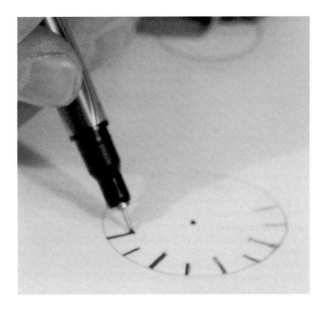

3. Ink your pencil lines using a fine-point permanent marker.

4. Add any numbers, calibrators, or text as neatly as possible.

5. Skip to Stage 2: step 3.

Hint

An alternative approach would be to draw out a much larger gauge, which would enable you to create much greater detail on the larger scale, and then reduce it using a photocopier or scanner, graphic software, and printer. For further details on how to do this, see step 2 under Method C, next.

Method C: Reproduce an Original You can actually reproduce an existing gauge face for an authentic appearance. In this case, I had an old gauge with a cracked housing. It might actually be desirable to have a gauge with a cracked housing—especially if it's a pressure gauge (indicating that at some stage the pressure reached critical levels). But the cracked gauge I had was of a less than useful size, so I opted to disassemble it and reproduce the faceplate.

Hint

If you do not have an original gauge (or don't want to risk destroying one that you do have), you can find copyright-free images of gauges online (or from the images in this book) and scale them to different sizes before printing them out.

1. I removed the faceplate from the cracked housing. I used white tape (white-out or white paint would also work) to mask off anything on the original that I did not care to reproduce—in this case, the name of the gauge manufacturer.

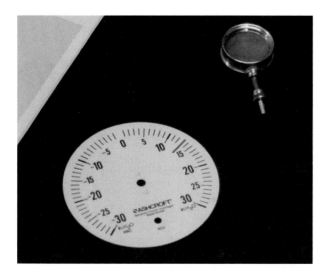

If I wanted to replace the gauge's text with text of my own devising (such as the laboratory or armory for the device of which the gauge will eventually be a part), now would be the time to add it. It will be easier to add this to the larger source gauge face at this stage than to the significantly smaller reproduction. In my case, I decided that this gauge will eventually be part of some sort of time-travel device, so have replaced the manufacturer's label with the words "Temporal Distortion".

2. Next I used a photocopier to reproduce the plate, manipulating the percentage controls on the photocopier until I had copies of the faceplate in various sizes (applicable to the sizes of the gauges I was constructing). Glossy photo stock or even coated mailing labels work well for this purpose.

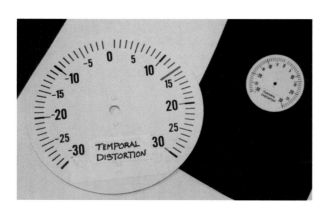

Hint

When using a photocopier or scanner, place a piece of black paper or cloth behind the source object to help ensure ideal contrast settings.

The process for determining the correct size is as follows: Measure the size of the faceplate (let's call this the source plate or "S") and divide it by the interior measurement of the cap/cup (this is the size of the faceplate for our project; let's call this "P"). Multiply that number by 100 to get the relevant percentage (S/P × 100). In my case, P needed to be 33 percent of S. Then carefully cut out your copy (P) and check the size against your cup/cap housing.

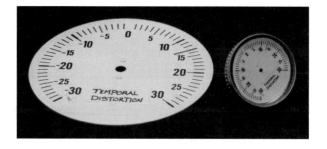

Note

I could also have chosen to scan and scale the faceplate using my computer scanner graphic software and printer. Your process will, of course, be dependent on whatever equipment you have accessible.

Step 3 After you're satisfied with your gauge face, you can add color. Colored gauges always read better than blank black-and-white gauges, but this really comes down to personal preference and design choices. To color in my gauges, I use red, yellow, and green markers, though I have found that translucent paints and even highlighter pens work well for this purpose. You can give your gauge full color, or adding just a "red zone" (as in a danger zone) can make your gauges appear more effective.

Note

If you are using computer-aided design to create your gauge, and you have a color printer, you can obviously perform the coloration of your faceplate digitally before printing it out. To so do, follow the instructions or tutorials that came with your software.

Step 4   Finally, we can affix an indicator to our gauge face—the arm/pointer that shows a reading on the scale of the gauge. As with everything else, we can do this in a number of ways.

Hint

Design-wise, I find that it is always preferable to position the indicator in the red zone or danger zone of a gauge scale. It gives the impression that something serious is happening, it creates a level of tension (no matter how small or insignificant), and frankly it just looks kinda cool.

- Using a store-bought pointer or reclaimed clock hand, you can rivet the hand through a hole punched through the center of the gauge face using a small double-capped rivet.

- You could use a small paper fastener, such as those sold by scrap-booking suppliers, as the pointer.

- You could glue an indicator hand in place. Because these gauges are purely ornamental and do not actually measure anything, the position of the indicator can be predetermined and set in place.

- If you want, and you are feeling kinda lazy, you could simply draw the indicator onto the gauge face using a straight edge and a sharp marker. This might be especially preferable with the smallest size gauges, where finding a small enough indicator hand could be difficult.

Hint

If you do not have (and cannot find) an indicator hand, you can make one by drawing out your shape onto a piece of black card stock or thin brass sheet metal and then carefully cutting it out before fixing it in place using one of the methods described. Your decisions concerning your design at this stage of the project will be entirely dependent on the practicalities of scale based on the size of the gauges you are making.

Stage 3: Enclose the Gauge Face Within the Housing

Now we can enclose our gauge face within the housing. At the stage, the various parts really start to resemble a functional gauge.

Step 1   Begin by using a small amount of glue (such as white glue or spray adhesive) to adhere your gauge faceplates onto a piece of dense cardboard. For this purpose, the back of a standard spiral-bound notebook would work, or you can work with thicker cardboard if you prefer. Generally speaking, the larger the diameter of your gauge, the thicker your cardboard should be (for strength and rigidity). Then, using a good pair of scissors or a craft knife, carefully cut out these circles to serve as backing plates for your gauge face.

Hint

If you want your gauges to look old, aged, and well-worn, consider taking a light colored furniture touch-up marker (or any light brown marker) and add a bit of "ink" on your fingertip; then rub the faceplate with your finger lightly to "age" the faceplate and make it somewhat grubby looking. For those of you skilled with graphic manipulation software, this could, of course, be done digitally prior to printout if you have chosen to use Method A: Computer-aided Design.

Step 2   Check to ensure the tight fit of the backing plates within the gauge housing.

Step 3   Pack some modeling clay (available from most hobby stores or craft shops) into the area inside the gauge housing up to the level where it will meet the backing plate. Leave enough room for a) the backing plate, b) the gauge faceplate,

and c) the clear front cover (see Stage 4). To ensure that the surface of the modeling clay is smooth and flat, I use the head of a large nail as a tamper to pack the clay into place.

Note

As an alternative to modeling clay, you could use any sort of putty—even silly putty could work for this purpose. But in my experience, modeling clay and/or epoxy putty are best.

Step 4   Fix the faceplate (Stage 2: step 2), the indicator hand (Stage 2: step 4), and the backing plate (Stage 3: step 1) together using a rivet, fastener, or glue. I will refer to this henceforth as the "faceplate assembly."

Step 5   Insert the faceplate assembly into the gauge housing, as shown in the next image.

Note

Leave a gap or depression above the faceplate assembly and below the rim of your gauge housing. The size of this gap will depend on which method you choose to construct your clear front cover (see Stage 4, next). If the gap is too large, fill the gauge housing with a bit more clay or with a second backing plate. If it's too small, remove some of the clay.

Stage 4: Clear the Front Cover

We have two main options for constructing the clear front cover for our assembled gauge. Your choice of Method A or B will depend on the scale and size of your gauge as well as the tools and materials at your disposal.

Method A: Liquid Cover, Two-Part Acrylic Epoxy

When making small gauges, this method seems to work best. It requires no special tools or skills, but it does require a steady hand and a degree of patience. If you do not think you can be patient, I suggest you explore Method B.

Step A-1   Mix up some two-part acrylic epoxy, which should be available from any decent hardware store or craft center. If you cannot find it locally, it is readily available online.

Note

You should always read all the instructions on your acrylic packaging, but in case you don't bother, here's how to use this stuff: Mix up the two parts on a palette (this should be disposable, such as a scrap of cardboard or an old plastic or glass bowl that you do not need for anything else...ever) at a 50/50 ratio and stir it with a stick for about one minute.

Hint

When portioning out your two-part acrylic, you will want a puddle roughly the same size as your gauge housing diameter for each of the two parts—that is, if your gauge housing is the size of a quarter, you will want a puddle about the size of a quarter of epoxy part A and another quarter-sized puddle of epoxy part B. Mix them together thoroughly before use.

Step A-2   Using a disposable craft stick, popsicle stick, chopstick, or similar, drop and dab the acrylic onto the faceplate assembly (already installed within your gauge housing), and allow it to pool until it fills the gap between the face of the faceplate and the top edge of the housing rim. See the image at the top of the next page.

Hint

It is often handy to tap the back side of your gauge housing gently to release any bubbles before the epoxy becomes too set or solidified.

Step A-3   Level off the epoxy carefully using a disposable straight edge (such as a scrap of cardboard, a playing card, or a popsicle stick). Wipe any overflow from the rim of the gauge housing.

Step A-4   Set the filled gauge aside and leave it alone for at least 24 hours before handling it. Be patient; handle the gauge before the epoxy is

set and you could run the risk of fingerprints or small bubbles developing within the clear coat (thus ruining the effect).

Method B: Rigid Cover, Clear Acrylic or Plexiglas

Assuming you have at least a moderately well-equipped workshop or evil genius laboratory, this option should not prove too difficult and is perhaps the most versatile and durable of the methods described here for constructing a clear front cover on your gauge.

Step B-1   You should already know the size of your internal diameter for the gauge housing—we have, after all, used this measurement several times thus far. Proceed, therefore, to mark out a circle on a thin sheet of clear acrylic or Plexiglas the same diameter as your internal gauge housing. I would recommend plastic of at least 1/16-inch thick, as the recessed/depressed gauge face gives a more realistic impression. Remember that in a

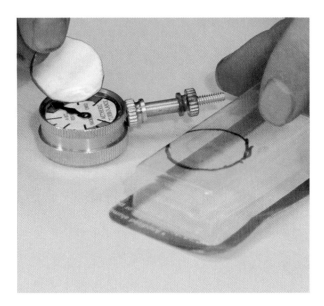

might want to use a band saw or a rotary cutting tool. Even a fine-bladed coping saw can be useful in this regard. I happen to have an industrial cutting laser, which makes the task quick and effortless, but I cannot assume that every reader of this book has a death ray hiding in the garage—then again, maybe you do.

Note

For further information on cutting out these discs, please refer to the instructions for cutting out the goggle lenses in Chapter 5.

real—functional—gauge, a thin space would exist between the clear front cover and the faceplate in which the indicator hand could move freely. If your faceplate is flush with the top of your housing, protected only by an ultra thin film of clear cover, it will not create the same impression.

Step B-3   After you have cut out your clear discs using whatever means is most convenient or accessible to you, dab a very small amount of clear glue (avoid super glue as it tends to cloud the lens) around the edge of the rim and pop the clear cover lens into the housing.

Note

If you are making very small gauges—say, ½ inch or less in diameter—you might well be able to get away with a flush mount. In this case, you can use thinner plastic such as is often found on blister packaging or food packaging. But, frankly, on the smaller gauges, I would recommend that you use the liquid epoxy of Method A.

Step B-2   After you have marked out your clear cover circles on your acrylic or Plexiglas, proceed to cut out these clear discs carefully, using whatever tool you have on hand that is best suited to the purpose. For very thin material, scissors or a craft knife can suffice. For thicker material, you

Conclusion

That is it then. It is that simple, that straightforward. If you have followed along successfully though the entirety of this chapter, you should have at least one or hopefully more gauges that look something like those shown at the end of step B-3. As noted earlier, these are merely examples of the sorts of gauges that can be made to add adornment, decoration, functional aesthetic, and verisimilitude to your other Steampunk projects, as shown in the images below.

Feel free to experiment and explore. Develop the idea and see what you can do with it. And if you see me (or Lord Featherstone) out and about somewhere and anywhen, do take a moment to show me what you have made with this and other designs within the book. Have fun with it and good luck.

See Chapter 5.

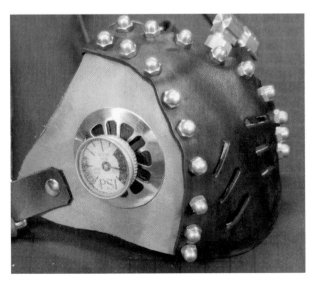

See Chapter 10.

See Chapter 12.

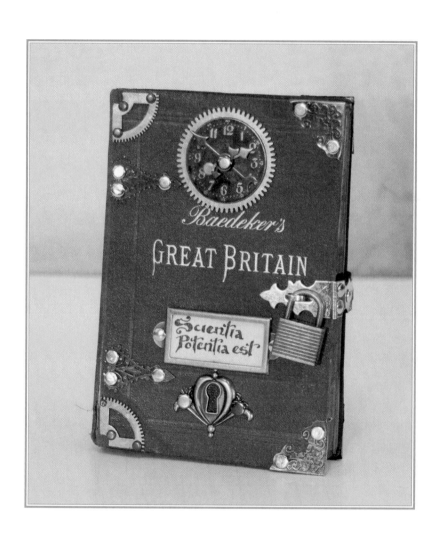

Chapter 7

Professor Grimmelore's Ferromagnetic Self-Scribing Automated Encyclopedia (or, The Steampunk Book Drive)

"What does the bloody thing say?" insisted Featherstone, a mere four seconds after Professor Grimmelore had finished writing his question in the book. He had just closed the cover and could feel the book vibrating with the movement of the tiny gears and push rods.

"Archibald, will you please be patient?" He knew it was useless before he was done saying it. Lord Featherstone was not on the best of terms with patience. He might as well have been asking him to stay sober.

"First names is it? That means it's bad news, Adrian old boy. You always use first names when giving bad news." The Professor ignored him for a moment. The book was still clicking and whirring. He imagined seeing the tiny magnetic arms rearranging the ferromagnetic ink to form an answer to his query. They needed to know exactly when the *Alexia* sank off the western coast of Italy on the second of August, 1743. Seeing as they both happened to be aboard the *Alexia* and it was the second of August, 1743, this qualified as a bit of a pressing matter. The book stopped whirring, and Professor Grimmelore opened it and read the entry.

According to the records of the Molo San Vincenzo Head lighthouse in the Gulf of Naples, Italy, the *Alexia* foundered at 15:23 local time. At 15:48 she fired off a distress shot indicating the ship was sinking. The cargo of fine glassware and all crew were lost.

"Featherstone, what time is it?" His Lordship glanced at the watch built into his mechanical arm.

"It's just gone a bit past two. Why do you ask? What does it say?"

"It says we need to get of this boat in the next hour and twenty minutes. Are you sure your watch is correct?"

"Of course it is. I set it at the observatory in Oxford only two days ago."

"Oxford! Bloody hell. We must get off this boat now!"

"What are you on about? Just because you are a Cambridge man doesn't…."

"The only problem I'm having with Oxford at the moment is that it is an hour in the wrong direction. We need to leave now." Adrian interrupted, pulling Feathers toward the ship's rail. Feathers pulled up short and pointed out into the sea at a bright light headed toward the ship beneath the water.

"What is that?"

"I honestly have no idea, but I bet it will hit this boat very soon causing it to founder at three twenty-three this afternoon," said the Professor.

"Adrian, um, you do know that I can't swim, don't you?"

"Oh... Bugger...."

—From *Featherstone and Other Unnatural Disasters:*
My Travels with a Mad Man, by Professor Adrian Grimmelore

*B*ooks used to be one of the best permanent depositories of knowledge. They were the equivalent of hard drives for hundreds of years. So wouldn't it be great to use to book to hold a modern hard drive? It would look much better on your desk than that ugly chunk of black plastic. Imagine your professor's face when you hand it to him for your PowerPoint presentation. It would also add a level of security to your data, because nobody would think of looking in that old book for your data.

Project Description

In this chapter we are going to examine a fun little project whereby we will modify an old unwanted book (preferably leather bound and Victorian, with rigid covers and a decent spine) to house a modern portable external hard drive.

Note

Read through the entire chapter before you begin working on this project, because what is presented here is just one possible variation. As your design might vary from mine, you might require different tools and materials to those listed here.

What You'll Need

To complete the project as described in this chapter, you will need to get your hands on the following materials and tools.

Materials

- Small external hard drive
- Piece of wood the same size and thickness as the external hard drive
- Hardcover book
- Wax paper or aluminum foil
- White glue
- Approximately 1 square foot of 1/4-inch plywood
- Assorted hardware for ornament and book furniture
- Paper fasteners
- Superglue
- Double-cap brass rivets
- Light belt-grade leather
- Hasp
- 6–12 small brass wire brads

Tools

- Safety glasses
- Marker and/or pen and pencil
- Band saw (or handsaw or power saw)
- Medium-grit sandpaper
- Glue brush
- Adjustable wood clamps or vise
- Wood scraps
- Power drill or brace and bit
- 3/8-inch spade drill bit

- ✐ Reciprocating saw or coping saw
- ✐ Craft knife
- ✐ Small metal file or grinder
- ✐ Center punch
- ✐ 1/16-inch drill bit
- ✐ Needle-nose pliers
- ✐ Rotary hole punch
- ✐ 1/8-inch drill bit
- ✐ Metal sheers or aviation snips
- ✐ 1/2-inch spade drill bit

Stage 1: Create a Wooden Hard Drive Stand-in

The physical dimensions of your hard drive will determine the size and suitability of any given book for this project. This example features a 160GB external hard drive by SimpleTech, chosen because of its plug-and-play features and compact dimensions. The drive measures 3/4 inch × 3 1/8 inches × 4 7/8 inches.

Step 1   After you've measured your hard drive, you'll need to create a wooden stand-in to take the place of the hard drive throughout the construction process. You don't want to risk beating the actual hard drive to death as you work on the project. We select, therefore, a scrap of wood that is 3/4 inch thick and trace out the external dimensions of the drive.

Put on your safety glasses. Cut out the stand-in with a band saw with a wood-cutting blade. Of course, if you do not have access to a band saw, you could use any sort of power saw, such as a table saw, circular saw, chop saw, or reciprocating saber saw, or even a good old fashioned handsaw. Lightly sand the edges as necessary.

Hint

When cutting out your stand-in, err on the side of caution. It is better that your wooden stand-in be slightly larger than the actual drive than smaller, because the rest of the project will effectively be shaped around it.

Step 2 Because this hard drive connects to a PC via a USB cable (rather than wireless), we mark the position of the USB port on the wooden stand-in.

Stage 2: Select Your Book

Step 1 I have assembled a small selection of suitable books that I have collected at flea markets, charity shops, the homes of deceased relatives, and other similar venues. From the books I had available, I selected a 1906 copy of *Baedeker's Great Britain*. Your selection can be any hardcover book that looks good to you, so long as the dimensions are suitable for housing your hard drive.

Note

Do be certain to investigate the value of any antique book before chopping into it.

Step 2 Read the book. *Seriously*. It is likely to be your last chance. If you are not interested in reading the book (or it is written in a language that you cannot read), at least page through it and cut out any maps, illustrations, or engravings that you think might be worth saving. In fact, being a travel guide, this book was chockfull of potentially useful maps, which I saved. After you have considered the important parts of the book, proceed to Stage 3.

Stage 3: Change Your Book into a Block

Step 1 We now proceed to glue all the pages of the book together. Insert a piece of wax paper (or aluminum foil) four or five pages from the end of the book. Make sure the wax paper is tightly wedged into the depth of the spine and that edges extend at least 1/2 inch wider than the pages.

Note

My book has two ribbon markers. I could, of course, cut these off, but I have chosen to leave them protruding from the bottom edge of my book, because they serve to emphasize that it is (or at least was) an actual book and not some sort of odd box.

Step 2 In a disposable bowl or container, combine about 1/2 cup of white glue and 1/2 cup water (a 50/50 ratio, in other words) and mix thoroughly.

Note

Mixing the glue with the water helps the adhesive work its way into and between the pages. And because the glue will not adhere to the waxed paper, it helps to keep the glue from spreading onto parts of the book where we do not yet want any.

Step 3 Open the first five or six pages of the book, and leave these open toward the book's front cover. Using a glue brush (or disposable paint brush), paint the edges of the book's remaining pages with a thick coat of the glue/water mixture.

Step 4 Insert a second sheet of waxed paper on top of the pages you pasted, between those pages and the five or six pages nearest the front cover of the book. Then close the book.

Step 5 Clamp your book tightly shut using a vise, or use a couple of boards and some wood clamps (or even a heavy weight). I happen to have a bookbinder's vise, which is rather ideally suited to the task. It is one of those items in my workshop which I am uncertain about how, when, where, or even why I have it. The first time I made one of these covers, I looked all over my shop for a suitable solution for clamping the book; when I stumbled

upon this bookbinder's vise under a workbench and thought, well, now, isn't that convenient.

Step 6   The glue should take about 45 minutes to dry. While your book is still tightly clamped, apply multiple coats of glue to the edges of those pages between the wax paper sheets. Then leave it to dry thoroughly overnight.

Note

Do not dispose of the remaining glue mixture, because we will be using it again later in the project. But do cover it tightly to be certain that it will not dry out overnight while you are waiting for the glued pages to set.

Stage 4: From Block to Box

Step 1   The glue on the edges of the book should be dry before you proceed. Remove the book from the vise or weight, and remove the wax paper you inserted between the pages.

Step 2   Place the book onto a piece of 1/4-inch plywood, and trace the outline of the book—including the cover—as shown here. If your plywood has square corners, save yourself some cutting work by positioning the book along an existing corner.

Note

If you do not have any 1/4-inch plywood on hand, you could use MDF or fiberboard, or you could cut the panels out of an old clipboard. Do not, however, use something like cardboard or foam core, because the following stages will require a material with greater rigidity.

Step 3   Repeat step 2 and trace another outline profile on the 1/4-inch plywood.

Step 4   Put on your safety glasses. Using either a standard handsaw or power saw, cut out both pieces of plywood. You now have two pieces of wood that are the same size as the outer dimensions of your book, in addition to the wooden book blank stand-in that you made in Stage 2.

Note

Lightly sand edges of the plywood if necessary.

Step 5   Fold back both of the book's covers and the loose pages of the book, so that the glued-together "block" section of the book rests flat on your work surface. Yes, I am aware that this is exactly what our school teachers told us never to do to a fine book, but, then again, they probably wouldn't approve of drilling and cutting into it either. Teachers and bibliophiles would be well advised to look away now lest you become permanently and irrevocably scarred by the literary desecration that follows.

Caution

Be sure that both the front and back covers are folded back and away from the glued-together block pages. You do not want to drill or make cuts through the cover.

Step 6   Position your stand-in blank on the glued block section of the book so that it is centered and equidistant from all edges. Then carefully measure the distance between the edge of the stand-in block and the interior spine of the book. Remember this; you will need this measurement in step 8.

Step 7   Sandwich the glued pages between the 1/4-inch plywood panels that you cut in step 4.

Note

Be sure to align the edges of both wood panels tight against the book's interior spine.

Step 8   On the innermost loose, unglued page, find and mark a "registration mark" at the exact vertical center of the book. Make a

corresponding mark on the vertical center of the top plywood panel as well as your stand-in blank. Then mark out your edge allowance (the distance between the interior spine of the book and the innermost edge of the stand-in blank, from step 6) on the wooden plywood panel—in our case, it measured 3/8 inch.

Note

The purpose of these registration marks is to ensure correct alignment as we assemble, disassemble, and reassemble our book drive.

Step 9   Trace the position of the stand-in blank onto the top plywood panel, and then remove the blank. Tightly clamp your plywood-panel-and-page-block sandwich together using at least two clamps (you will remove the clamps one at a time to make some cuts and want the sandwich held tightly throughout that process). Make sure the registration mark on the plywood aligns with the mark you made on the book page, as shown on the top of the next page.

Hint

I have used a clothes pin (though a large paper clip, small spring clamp, or even a large rubber band might work as well) to hold the book covers and loose pages together and out of my way while I work on the page block section.

Step 10   Put on your safety glasses. Then grab your drill and a spade bit. We'll drill two holes in opposite corners of the plywood panel marked with the stand-in blank's profile. Choose a spade bit that is slightly larger than whatever saw blade you will be using to cut through the glued pages. In my case, I will be using a reciprocating electric power saw with a fine-toothed blade (long enough to reach through the depth of the page/panel sandwich) measuring 1/4 inch wide. I have selected, therefore, a 3/8-inch spade bit, which should allow plenty of space for me to insert the saw blade after the hole has been drilled through the book and plywood panel.

Note

Use extra care when drilling out these holes, and be certain that your "sandwich" is firmly secured to your work surface.

Step 11   Place a scrap of wood under the entire project, and clamp the wood scrap, plus the entire plywood-sandwiched book block, to your work surface. Then carefully drill holes through the top plywood piece, the glued page blocks, and the bottom plywood piece.

Caution

When drilling out these holes, proceed slowly and drill in short bursts if you can. A tremendous amount of friction and heat is generated during this process, and that paper dust you are creating as you drill through the glued pages can be highly combustible. It would be a shame to set your whole project on fire at this stage.

Hint

Use a piece of tape to indicate a drill depth on your drill bit, to avoid drilling too deep and damaging your work surface. This depth will let the drill pass through the page/panel sandwich and into the scrap of wood beneath.

Step 12   Now you'll cut out the hollow space in our book where the drive will be concealed. You can use a small deep-throat coping saw to make

this cut by slotting its blade through your pilot holes and then beginning your cut. Or you can use a reciprocating power saw to make the cut. Keeping the sandwich firmly clamped together, cut along the lines by slotting the saw blade through the pilot holes you just drilled. You might need to adjust one of the clamps to make room to work.

Note

If you are new to using a coping saw, refer to instructions provided in Chapter 9.

Step 13   While your book sandwich is still tightly clamped together, move a clamp to make room to work and finish cutting around the edges of the traced areas. Ensure that all of your internal edges are straight and your internal corners are clean and square.

Step 14   After you have finished cutting all four edges, gently push the panels and center of the pages out of the book. Remove the clamps and the facing plates from your book sandwich. You should now have a large rectangular hole in the center of your book. Now, using your stand-in blank, test that the size of the opening you just cut fits your hard drive to hold it snugly in place. If your cut lines have wavered at all, trim off any excess with a sharp box cutter or craft knife.

Stage 5: Create the Bottom of the Box

Step 1   Retrieve the unused water/glue mixture, which I am sure you saved as previously instructed (or mix up a fresh batch if you didn't save it).

Step 2   Using a glue brush, apply glue to the inside of the back cover of the book and glue the last page to the back cover.

Step 3   Glue together any loose pages at the back of the book. Proceed systematically, gluing one page at a time, until all of the pages at the back of the book are firmly affixed to each other and to the back cover.

Step 4   Insert a piece of wax paper between the cutout section and the loose pages at the front of the book. Make sure the edges of the wax paper extend at least 1/2 inch beyond the edges of the book's pages. Lay the book on a flat surface face (front cover side) down.

Step 5   Thoroughly coat the edges of the back of the cutout and all of the interior walls of the hard drive hollow with glue. You will be gluing the back pages (with the cover) to the back of the cutout section, but while the book is still open, you want to bind the inside of the hollow with glue in the same way you bound together the outside edges of the pages in Stage 3.

Step 6   Now clamp the book tightly together (being certain to leave the wax paper sheet in place) and allow it to dry completely overnight.

Step 7   Set the clamped assembly aside and switch your attention to the book clasp and *furniture* (the metal mounts on the outside of antique book covers).

Tip

Any good Steampunk maker worth his or her steam will be something of a packrat (or the term we use at Brute Force Studios, "brassrat") when it comes to interesting parts and components. Save everything despite your having no idea what you might actually get around to using it for.

Stage 6: Add the Book Clasp and Furniture

When we make these book drives, no two ever end up being quite the same. One reason is the specific characteristics of the book, which gives its life to be repurposed, and the other reason is the assorted hardware that becomes part of the ornamentation and furniture for the book. I usually dump out a bunch of suitable hardware onto my work surface to survey for shapes, designs, and general inspiration. I find some of these items at large craft stores that sell items for jewelry-making and scrapbooking. (Yes, I said it—scrapbooking supplies. I promise to gargle with liquid soap just as soon as I finish writing this chapter.) Other items come from antique shops, charity shops, or flea markets. You can often find a "box of old hardware," full of rusted screws or bolts and the like, which will likely contain at least one or two tasty little interesting bits of old furniture hardware. If the box full of junk can be picked up for a few dollars, those one or two interesting pieces might well be worth the purchase price, because buying them individually could be far more expensive.

Tip

If you have another book of roughly the same size as your project book, you can use it to proceed to lay out all of the book furniture while the primary project is waiting for its latest glue to dry. If you do not have another book of comparable size, you will have to wait until the glue has dried completely before working with the book.

Step 1   Lay out the project's hardware/ furniture components. What I've chosen is shown in the image at the top of the next page.

Step 2   After you've gathered all the pieces to use on your book cover, place them on the cover to check their position and see if you have everything you need. And, as luck would have it, I do.

Note

Of course, your project is unlikely to be identical to this one, but let us assume from this point forward, to make this process a bit easier to explain, that your book is the same as my book. Your hard drive is the same as my hard drive. And all of your furniture and ornament is identical to mine.

Step 3   Now we can attach the hardware. We'll start by attaching the two front corner pieces that I cut out of my brass gears (see the sidebar).

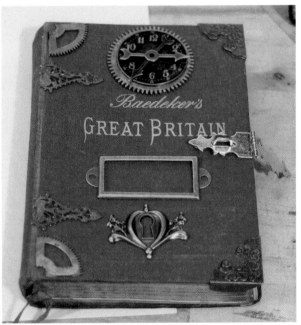

As often happens with these homemade corners, these pieces are not quite large enough to

Modifying Clock Gear Corners

Most of the pieces I have chosen are suitable as they are, but I want to construct some of my corner pieces out of some old salvaged (that is, pillaged) brass clock gears. Because this is kind of tricky to explain in words, please follow along by looking at the photos.

I start by marking the gears based on how I want them to appear on the book cover's corners.

I am attempting to get two corners out of each gear. I could carefully cut four corners from each gear,

especially because these are four-spoke gears, but when I have done that in the past, I have found the proportions to be aesthetically unpleasant. I am trying not to be greedy and going for two good corners, rather than four mediocre corners. Trust me; it will work better that way.

I cut the corners out of the gears using a band saw, but I could also use metal shears or even a hacksaw. Once cut, I grind or file off any sharp edges (but not all of the coggy teeth, obviously) using medium-grit sandpaper, a small metal file, or a grinder.

accommodate a double-cap rivet (which we'll use to attach the other elements). So we will attach these in a unique way.

Step 3a Using a fine-point marker, mark the positions on the brass pieces for three holes—one in each corner and one in the middle along the shortest corner. Use a center punch (as discussed in the goggle project in Chapter 5) to dimple the holes on the brass pieces.

Step 3b Clamp the brass pieces to scrap wood on your work table (or secure as shown in the photo) and use a 1/16-inch drill bit in a power drill to drill out the holes on each of the brass pieces. Repeat this process for four book corner pieces (two on the front and two on the back).

Step 3c   Now move to the book covers. Position two of the corners on the front cover—on the two inside corners (the corners adjacent to the spine)—being sure that they are tight into the corners of the cover, yet do not interfere with the normal operation of the book spine. Proceed to mark the position of the holes on the corner gear pieces onto the book cover and then drill (or punch) them out.

Note

Because these holes have been marked by hand (and they are likely not mathematically precise), be certain that you keep track of which side of the corners is up and which is the top or bottom corner.

Step 3d   Attach the brass gear corners to the book front cover using superglue and some small paper fastener type attachments that came packaged with some of the other furniture hardware.

Note

These little metal paper fasteners are available separately from most places that sell scrapbooking supplies. Alternatively, you could use small pieces of wire or even micro-rivets to affix these corners in place.

Step 4   To attach the opposite (outside) corners, we'll use two brass corners that are

commercially available through scrapbooking suppliers (yes, there is that dirty word again). Position them and then use needle-nose pliers to bend the flaps around the cover's edges.

These corners have a tendency to come off, so secure them in place by marking, dimpling, and then drilling out holes for two small brass double-cap rivets. Repeat this process for the other (last) corner on the front cover.

Note

Be sure to set the rivets in far enough on the corners so that the rim of the rivet cap will not overlap the edge of the corner piece. For more details on riveting, see Chapter 5.

Step 5 Now it is time to get fancy. Mark and position the rest of your furniture on the front cover, being certain to leave space for the strap/book clasp. Then attach the furniture in whatever way seems most appropriate to the hardware you have chosen—you can use rivets, paper fasteners, superglue, or a combination.

Hint

Because one of the pieces I have chosen to include as furniture on this project is an ornamental keyhole, it is probably worth mentioning that the keyhole illusion is far more effective if the space behind it is colored in black (suggesting a vacant space behind).

Step 6 Next we turned our attention to the strap/book clasp/hasp. The leather strap and hasp will keep the book "closed." Attach the hasp loop to the middle edge of the front cover using small double-cap brass rivets, similar to how we attached the other furniture onto the front cover. (Noting that my hasp end was longer than the thickness of the book, I trimmed off the excess and drilled a new mounting hole.)

Step 7 Cut a strap of light belt grade leather (this can come from an old belt or even from a discarded purse) to the same width as the hasp (in this case, just over 1/2 inch) and about 4 inches long. Attach the hasp onto the strap with double-cap rivets. It should look something like this:

Step 8 Slot the hasp over the loop and hold it in place using a scrap of wire (a small nail or even matchstick or toothpick would also work). Then stretch the strap around to the back cover and decide where you want to attach the end. (I attached it at the center of the back cover.)

Step 9   Using a rotary punch, punch a hole near the end of the strap where it will be attached to the back of the book. Then trim off the end of the strap to leave about 1/2 inch beyond the hole. Mark the position of the strap hole on the back cover, and drill straight through into the interior of the hollowed-out part of the book using an 1/8-inch drill bit. Use a rivet to attach the strap to the back cover of the book.

Note

In this case it is far easier to place the post through from the outside and the cap on the interior. Hammer flat.

Step 10   Once all of the furniture and ornament is safely attached and secured to the front cover and your hasp and strap are attached, you can turn your attention to finishing off the back cover. Begin by attaching the four corners (two cut from the brass gears and two scrapbook corners) onto the back cover. Because the back cover is glued to the pages, it presents a few different challenges, with some different attachment possibilities.

Step 11   Let's start with the scrapbook corners. Using metal shears or aviation snips, trim off the flanges that would otherwise be folded around the cover corners.

Note

If we had attached the corners prior to gluing together the cover and the pages, we might have left the flanges on the cover corners, but they might have interfered with the pages meeting flush and smooth to the back cover.

Step 12   Mark and drill two mounting holes through the trimmed corners using a 1/16-inch drill bit.

Step 13   After locating the correct position on the back cover, mark the holes and then drill them out about 1/2-inch deep using a 1/16-inch drill bit.

Use a dab of superglue in the underside of the corner piece and a dab of superglue in each hole before positioning the corner piece and slotting in two paper fasteners. Repeat for other corner.

Step 14 To finish off the back cover, we need to attach the spine-side corner brackets—the two remaining brackets that were cut from the brass cog. Position the corners and mark where the holes fall on the back cover. Drill out these holes using the 1/16-inch drill bit to a depth of about 1/2 inch. Then apply superglue to the back of the corner bracket and affix it in place by inserting small brass brads into the drilled holes. Add a dab of superglue on the shaft of the brad to hold it firmly and securely inside the hole.

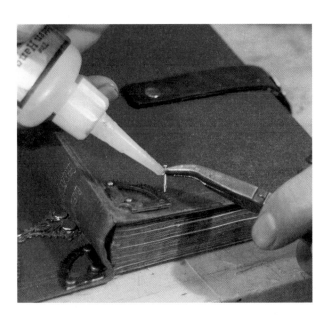

Note

Because you have drilled into the "meaty" bit of the book (which is well and truly stuck solid with glue, right?), you might even be able to get away with using some small screws to attach corner furniture to your back cover.

Stage 7: Add Access Ports

Now you need to examine your hard drive and determine the correct placement for any access ports—a USB port or power port, or both, or neither: A wireless drive will not require ports. A drive powered by a USB cable will require one port. A drive powered by a separate power cable that uses a USB will require two ports.

Step 1 Assuming you are using the same drive that I have, use the port position mark you made on your stand-in blank (Stage 1, step 2) to align and position the port hole on the edge of your pages. In our case, it is a single hole on the bottom edge of the pages.

Step 2   Put on your safety glasses. Clamp the book tightly to your work surface (or clamp it tightly in a vise). Drill out the port hole using a

1/2-inch spade bit in a power drill—1/2 inch is enough to accommodate the end of the USB cable that goes with the drive (the size of hole necessary to accommodate your port might differ, of course).

Step 3   The project is complete. Now fit the external hard drive into the interior of the book. The drive should fit in easily but snugly. If the drive is loose, consider lining the interior with thin suede or fabric.

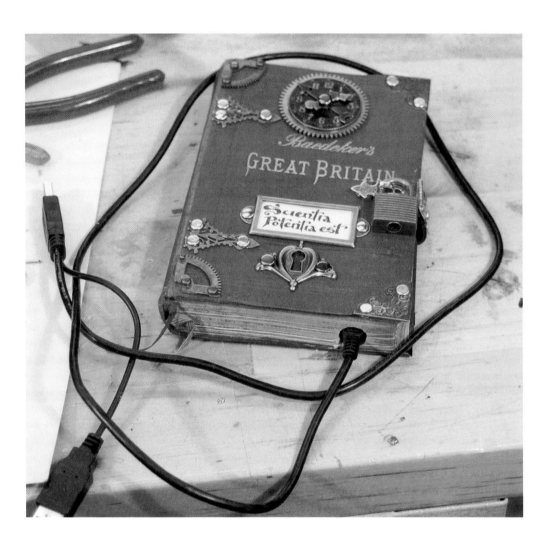

Chapter 8

Dr. Visbaun's High Voltage Electro-Static Hand Cannon (or, The Lamp Gun)

She could feel the weapon pulsing in her hand. When she had pulled it from the violin case, the gang of men laughed at her for brandishing such a delicate looking thing. When she wound the crank on the side and a crackle of electricity ran and jumped up and down the barrel, they stopped laughing, but they did not turn away. These men were obviously not yet familiar enough with scientific advancements to be as terrified as they should be. She would make one last stab at ending this in a civilized manner.

"Do not make me use this. Nobody needs to get hurt."

"Do you fink we is gon' turn 'bout and be on our way jus' cause you's pulled out a fancy torch?" one of the men said. The rest started to laugh a bit and regain their courage. They advanced a step and she retreated a step. She knew the only way out of this was through them. They knew it, too.

"I am giving you one last... OW!" One of the men had thrown a piece of brick and it had hit the wall next to her, spraying her with fragments. She could see the man being patted on the back by a few of the others.

Stuff civility. She leveled the weapon at him. Though it had a sight, she knew this thing was actually about as accurate as Dutch accounting. She had no idea whether it was going to hit him or someone/something else. When the bolt of lightning shot from the end of the electro-static hand cannon, all she knew was that someone over there was going to get hurt very badly. This made her smile.

—From *My Life in the Air*, by Ms. Adelaide Grayson
(Major, Her Majesty's Aero-Forces, Ret.)

You cannot go about having adventures and such without being suitably armed. Your weapons need not be lethal, but they must be effective. Outrageous weapons have been one of the defining points of most any type of science fiction. We Steampunks are no exception. I could show you how to modify an existing toy gun from We-Be-Toys and stuff. In fact, I have seen some incredible examples of such things in my travels. But in my less than humble opinion, they aren't a patch on a scratch-built piece of portable artillery.

Project Description

This project will detail just one of a virtually infinite number of ways to create a ray gun from various reclaimed or found parts. In this case, we will be turning parts of an old lamp into the deadliest weapon known to humanity—well, no, not really, but it should end up looking pretty cool.

What You Will Need

To complete the project as described in this chapter, you will need to get your hands on the following materials and tools.

Materials

- Lamp (or similar), preferably brass
- Decorative flintlock gun replica (or something similar to serve as the gun stock)
- 2, 3/4-inch, 8-32 brass machine screws
- 2 lamp finials
- Brass strip, 1-inch wide, 12-inches long
- 2, 2-inch, 6-32 brass bolts or machine screws
- 3, 6-32 brass nuts
- Hand crank (from clock, pepper mill, coffee grinder, hand drill, etc.)
- 3 brass cogs, or gears, in assorted sizes
- 1, 1 1/8-inch, 8-32 brass machine screw
- 6 small brass washers
- 3, 8-32 brass machine nuts
- 1, 8-32 knurled nut
- Small gear with a shaft
- Brass strip, 3/4-inch wide, 1/16-inch thick
- 3 or 4 #2 brass wood screws
- 3, 12-inches long by 1/8-inch diameter brass rods
- Copper wire
- 1, 1/2-inch, 6-32 brass machine bolt
- 1, 6-32 knurled nut
- Brass ornamentation

Tools

- Diagonal cutters (dykes)
- Pencil and/or marker
- Small ruler
- Safety glasses
- Band saw and wood-cutting blade
- Rotary tool with 1/2-inch sanding drum
- Clamps
- Awl
- Power drill
- 11/64-inch drill bit
- 1/8-inch drill bit
- Tap and die set
- Hacksaw
- Metal shears
- Fine-grit and medium-grit sandpaper
- 9/64-inch drill bit
- 2 sets of pliers
- Denatured alcohol
- Clean rags or paper towels
- Solder-It silver-bearing soldering compound
- Propane torch
- Soldering stand and screen
- Loctite adhesive or superglue
- 3/8-inch drill bit
- Bench vise
- Wood stain
- Disposable paint brush
- Polyurethane varnish
- Flat black spray paint

∽ Heavy-duty wire cutters, linesman pliers, or fencing pliers

∽ Spring-loaded center punch

∽ Phillips screwdriver (for mandrel)

∽ 1/16-inch drill bit

Stage 1: Select Your Primary Material (Lamp)

Select your lamp. Well, no, it doesn't necessarily have to be a lamp, but it should be a relatively long object, preferably made of brass, with a bulbous section (which would sort of indicate the gun breech's capacitor section, containing mysterious mechanisms, wondrous workings, or even magic crystals—who knows). Here's what I will be using, for reference.

If you do not have a lamp…um…get one. I mean you should start keeping a sharp eye open at flea markets, charity shops, yard sales, or wherever you can find bargains. Closely examine the shapes of any lamp on offer. Does it look like it wants to be something other than a lamp? I think so, too. And we are probably right. It probably does.

Caution

Do not be tempted to take your cheap old bargain lamp and plug it in to see if it works. Not, at least, before thoroughly inspecting every bit of the wiring. More often than not, vintage and antique lamps have seriously substandard wiring or even disconnected, frayed, or shorted out wiring.

Stage 2: Disassemble Your Lamp (or Whatever)

Step 1 If your lamp still has an electrical cord attached, cut off the wires using a pair of diagonal cutters.

Step 2 If your lamp has a heavy weight in the base (many do), remove it. These are usually

just threaded in place or held in place with a simple lamp nut attached to the threaded rod (often called an *all thread rod*) at the base of the lamp.

Step 3   Because I have no way of predicting exactly what sort of lamp you are using, all I can advise is that you continue to disassemble everything. Remove the socket assembly parts at the top of the lamp, where the bulb is screwed in. Remove the base of the lamp. Remove all bolts, screws, nuts, and fasteners until you are left with a pile of components. But most important, you should have a few yummy brass parts, the pieces that inspired you to turn the lamp into something else in the first place. Spread out your parts and components and survey them for inspiration. Which pieces will work well together?

Note

Let us assume that, for the remainder of our project, you are working with exactly the same lamp that I am using (which, of course, is not actually true). And let us further assume that your finished project will end up looking identical to mine (which again is highly improbable). Without these assumptions, it will be difficult if not impossible to guide you through this project. But please understand that the purpose of this particular project is to teach you how to use "found objects" and to see them for something they are not, and then to "upcycle" them into something entirely different—something never intended or imagined when the various components were made.

This chapter, therefore, is an example of this sort of project and is designed to share with you certain principles of "modding." Unlike many of the other projects in this book that create devices from raw materials by design, this project will be directed largely by whatever materials you might have on hand or that prove inspirational.

Step 4   Now you need to find something to serve as a gun stock, grip, or handle. Find something that it is structurally sound, because you'll be attaching all the other components to this piece. For this example, I have selected an old replica (that is, a decorative wall-hanging) flintlock pistol.

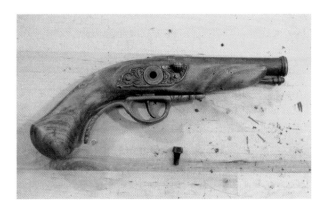

These decorative pistols are widely produced and can often be found at flea markets or online through auction sites such as eBay. Search for "flintlock," and you will see what I mean. The metal components on these decorative pistols are cast from cheap metal, as they are not intended for any sort of use and serve merely as decorative objects. These cheap castings break pretty easily, and for our purposes this is good, because broken decorative guns come cheaper than pristine examples. The main piece we want is the wooden gun stock/grip (to save us having to make one) and perhaps the trigger guard and trigger mechanism (or other useful parts). Disassemble all the pistol components by undoing any and all screws.

Note

Do not dispose of any of the metal pistol parts unless they seem broken beyond use; you do not yet know what sorts of pieces you might find useful.

Stage 3: Get It Together

Step 1   By now, you should have an idea of what parts of the lamp will serve as gun parts. You need to decide which part of the lamp will be attached to the gun stock so that you can marry up and attach the barrel parts to the gun stock.

Assuming, of course, that you are using the same parts and will be making a hand cannon identical to mine, start by tracing out the profile of the "breech" section of the barrel (the part attached to the stock) onto the gun stock. We'll be notching the gun stock so that this section of the lamp will fit tightly and securely into the stock. The part of the gun that will serve as the breech in this example is at the base end of the lamp I took apart, minus the actual base.

Step 2   Place the stock and barrel (lamp part) up tight against each other on a smooth and secure work surface, with the barrel above the gun stock. Use a small ruler and measure about 5/8 inch onto the body of the stock; then use a pencil to follow

and trace the contour of the barrel breech as far as necessary, as shown here.

Here, I've connected all of my measurements and outline of the breech and should, therefore, have a reasonable approximation of the profile of my barrel breech traced on to the stock.

Step 3   Having clearly marked out the breech profile in step 2, I don my safety glasses and carefully cut along the lines using a band saw with a wood-cutting blade installed.

Note

Because the butt of my gun stock is wider in the middle than along the sides, I place a shim under the stock so that the piece sits level on the band saw table and the blade will cut at the appropriate vertical angle. I've also used several vertical cuts into the stock to make it easier to maneuver the band saw blade through the piece.

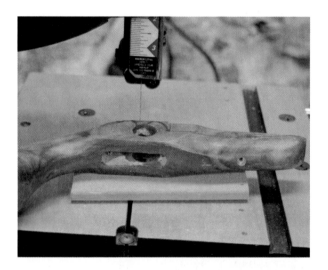

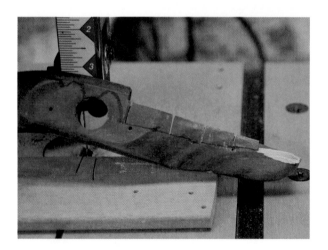

Caution

Many people will advise you to wear work gloves when using a band saw or table saw. I would strongly advise you not to do so. If your bare finger touches the blade it will, indeed, be horrible and possibly very injurious, but you are also likely to pull your finger away very quickly. If, on the other hand, your work glove gets caught up in the movement of the blade, it can often drag your hand forward, further into the blade, thus incurring an even greater injury.

Step 4   Once the primary cut is made using the band saw, make smaller, secondary cuts, also with the band saw, to keep forming the shape.

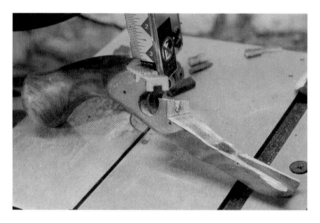

As you work, continue to match up the barrel breech to the cutout gun stock to check the fit. It is important to get a close fit at this stage to avoid problems later in the assembly of the project. If the fit is not precise enough, carefully trim or sand the cut-away area to shape.

Further improve the cutout shape in the stock using a 1/2-inch sanding drum in a rotary tool.

Step 5 After you have established a clean fit between the gun stock and the breech of the barrel, match them up and clamp them together.

Note
If your project is using something other than a gun stock for the grip and body of the gun, you might well need to improvise this step. At the very least, you will need to locate and drill a hole to insert a bolt and attach the barrel.

Then examine the underside of the gun stock for a hole that can be used to secure a fastener.

Mark the position of the attachment hole onto the breech section of the barrel using a marker or by scribing the point with an awl. You need to secure the barrel to the stock using a bolt, but be aware that you probably will not be able to reach inside the breech to secure a nut. In this case, we will drill out the hole and then thread it with a tap (from a tap and die set), so that we can screw the bolt directly into the breech.

We'll use a 3/4-inch long, 8-32 brass machine screw, which is long enough to slot through the gun stock and extend securely into the breech. With your safety glasses on, drill out the hole in the gun stock with a power drill fitted with a 11/64-inch drill bit.

Note
Some inexperienced, or lazy, "modders" might be content with using some sort of adhesive to secure these two pieces together (or even, god forbid, using hot glue), which might of course work for a while, and might look great. But the first time it is dropped on the ground (usually by an innocent, yet intoxicated, bystander), it is likely to break. And you wouldn't want that to happen after all of your diligent and careful work. Furthermore, as with all of our projects, we want, so far as is possible and practical, to be able to assemble, disassemble, and reassemble the project as many times as necessary. Rule number 1 of Steampunk making and modding is that screwed is always better than glued—make it your new mantra.

Step 6 Use a 1/8-inch drill bit to bore out the corresponding hole into the breech. Then tap out the hole, cutting a thread into it using my 8-32 tap, as shown next.

Note

This process will work properly only if the walls of your breech (lamp) are thick enough to accommodate a decent thread. If the lamp (or whatever) you have chosen is flimsy or thin-walled, tapping out the hole with a thread would be almost irrelevant.

If the walls of the breech are not thick enough, you should use a long 11/64-inch drill bit and drill all the way through both sides of the breech. Then use a long 8-32 bolt to go through both the stock and the breech to attach them together. You can use this same method if you are not comfortable using the tap and die set.

Step 7   Bolt it all together to make sure you have a good, solid, secure fit.

Step 8   And now, unbolt it all, so we can continue to work on the gun stock.

Stage 4: Make It Wow

If you have managed to follow along successfully thus far, you should have something that looks rather gun-like and could well serve as a gun prop. But, let's face it, you would not be reading this book if you would be satisfied with such an anemic example of Steampunk firepower. So now we need to turn our attention to ornamentation.

You want the finished project to convey the idea that every part of it has a purpose and a function. Although the Victorian era was rife with extraneous ornament in terms of filigree and carving and decoration, that ornament did not extend to unnecessary mechanisms, or random gears, cogs, or electrical terminals. Think about how this device would actually operate—assuming, of course, that the breech actually did house mysterious mechanisms, wondrous workings, or magic crystals. The decisions you make regarding ornamentation and "steampunkery" from this point forward will be dependent on your conception of how this piece is *designed* to function.

One way to determine how your gun is going to "work" is to give it a sort of theme. Is your gun going to be a laser or ray gun? Will it shoot some sort of pellet, slug, or bullet? Is it some sort of customized flame thrower? Or, as in the case of the cannon we are building here, will it store up an electrical charge in a capacitor and then unleash the positively charged ions in lightning bolts of destruction? The choice is yours, of course, but whatever you decide should guide your design process.

This is the hardest part to explain in a step-by-step fashion. Look at your stuff. So far, all you have managed to do with this project is to attach a lamp to a broken pistol stock. But now you want to look for inspiration among the shapes for "attachments." The attachments will appear to be functional pieces designed to serve some sort of important function in the operation of your weapon. Later, you might choose to add ornamentation in a Victorian vein, but for now look for functionality.

Step 1   Assuming that your "pile of stuff" consists of the same components as mine (which is of course rather unlikely), let's begin by fashioning a lamp finial as a terminus for the breech of the gun. In our case, the bottom end of the lamp is the breech attached to the gun stock. It will not do to have an empty threaded rod at the back end of the breech, where the rod from the lamp had attached to the lamp base. Using a hacksaw, saw off all but about 1/2 inch of the all thread rod, leaving enough of the thread to screw on the finial.

Hint

I use a hacksaw in preference to my metal cutting band saw because the all thread rod is made of some sort of steel alloy. Cutting this can dull the fine teeth on my band saw blades, which are primarily tasked with cutting brass, which is much softer.

Because almost all lamps built from the mid-20th century onward use a universal thread, a standard lamp finial can be screwed on the end of the all thread rod. The pieces you removed from the lamp probably include a finial. And of course you saved all those lamp parts when you took it apart, right? If not, you will find a variety of lamp finials at any place that sells lighting (such as nationwide hardware and superstores). Place the finial on the all thread rod to check the fit, then remove it.

Step 2   Now we turn our attention to the section of the gun stock that has a gap, where the flintlock mechanism was once fitted into the wood. We want to cover this unsightly gap. We use a piece of 1-inch-wide brass stripping (available through places that sell K&S Engineering products, among other sources).

On your work surface, align the brass strip against the bottom side of the gun stock. Use a marker to mark out (freehand) the size and shape of a plate large enough to cover the gap.

Step 3   Cut out the cover plate using a band saw, a rotary tool, or metal shears. Then grind or sand the edges. Place the plate in position on the gun stock and check the fit in terms of size and shape. Mark and make any necessary alterations so that it fits the way you want.

Step 4   Position and mark a spot for two holes on the plate, where you will insert screws to attach the plate to the stock.

At this point, you could drill holes through the brass plate and then use wood screws to secure it to the wooden stock. But I prefer, instead, to drill all the way through the wooden stock and then

secure the plate using brass bolts that extend straight through from one side to the other. In my opinion, having two good solid brass plates on the sides of the gun stock not only serve to give the device an impression of solidity and symmetry, but also help to strengthen and stabilize the entire assembly. Also, if the bolts extend all the way through, you can use the ends as faux electrical terminals, which we will do a bit later in this chapter.

Drill out two holes in the brass side plate using a 9/64-inch drill bit, to accommodate 6-32 brass bolts or machine screws. Then mark the corresponding spots on the wooden gun stock and use the same bit to drill those holes.

Step 5   Now create a second brass side plate for the opposite side of the gun stock. Start by repeating step 2, whereby you sketch out the size and shape of the plate freehand. Alternatively, for perfect symmetry, you could trace and transfer the shape of the first plate onto the second.

After you have traced the profile of your plate onto the brass strip, cut it out. Then grind, file, or sand the edges smooth. Now mark the position of the bolt holes and drill them out using a 9/64-inch drill bit.

Note

Using a marker, be certain to write an "R" for the right and an "L" for the left on the underside of the corresponding side plates so you know which is the right plate and which is the left (assuming, of course, that yours are not perfectly symmetrical).

Stage 5: Install the Crank Assembly

I want to add a hand crank onto the side of my gun. This crank will be used to wind up an induction coil of some sort (mysteriously contained within the breech). Once wound, a strong electro-static charge will be stored within a capacitor (also mysteriously hidden within the breech) until the trigger is pulled, unleashing the charge as a miniature lightning bolt…of destruction.

Small hand cranks are commercially available from dealers that supply clock components, because they are designed for winding large parlor and hall clocks, such as grandfather clocks. Or you can find them elsewhere. The one I use for this project came off of a pepper mill or coffee grinder that I found at a local flea market.

Step 1   I want to secure the hand crank to the side plate of the gun. Any alterations and additions should be made to the side plate at this point, prior to assembly. Begin by drilling out the interior of the shaft of the hand crank using a 1/8-inch drill bit. Then select a #8-32 tap from the tap and die set and proceed to cut a thread into the hole of the shaft.

Step 2   Mark the position on the side plate where the hand crank will be attached. Drill out a hole using a 11/64-inch drill bit. Select two or three small cogs or gears (from the assortment of random bits and parts and pieces that you always have on hand). My gears came out of some old windup alarm clocks I found at a flea market.

I have chosen two gears (and I assume that you have chosen two identical gears). They have specific names and specific purposes, but that doesn't really concern us now. For sake of ease, let us refer to them simply as the large gear and the small gear. And for their purpose, all we need to ensure is that their teeth roughly match up.

Step 3   Don your safety glasses, and drill out the center hole in the large gear using a 11/64-inch drill bit. Note that the small gear has an even smaller diameter gear connected to it by a very small shaft.

Caution

As you drill out this center hole, hold onto your gear securely with a reliable pair of pliers, rather than holding the gear in your fingers. I have found through experience that sometimes when drilling out a gear, the drill bit can catch and bind in the hole it is creating. When this happens, that gear wants to spin along with the drill bit. The gear effectively becomes a saw blade, and if it is a saw blade that is being held between your fingers while spinning, you might well end up with both unpleasant and unhappy results.

Step 4   Place the large gear on the plate where you will be placing the crank handle and mark the plate through the center hole of the gear with a marker. Place the small gear on the plate in the desired place with the teeth meshing with the large gear. By pushing the small gear down hard on the plate, the small piece of shaft on the bottom of the small gear will mark the plate at the center of the small gear. Lift the gear away and enhance this mark with a marker. Use an 11/64-inch drill bit to drill the hole for the large gear and a 1/16-inch drill bit to drill the hole for the small one.

Step 5   We are now going to do some simple soldering. Start by thoroughly cleaning all parts you are going to solder, using denatured alcohol (available from any hardware store) and a clean rag or paper towel. It is important that you clean these parts before soldering to remove any dirt or oils that can impede the soldering process.

Caution

As any metalworker or welder will tell you, hot metal looks just the same as cold metal. Metal does not have to change color to be capable of inflicting serious burns. Leave your soldered piece alone for a few minutes until you are certain it has cooled down enough to handle. Or, better yet, use a pair of pliers to pick up the piece and run it under the tap or dip it in a bowl of water before touching it.

Step 6   We will solder the little gear into place using Solder-It silver-bearing soldering compound (available from most hardware stores) and a propane or butane torch. In the photo, you can see that I have placed my pieces on a soldering stand and screen—this helps to ensure that the heat isn't sucked away by another surface (such as an anvil) and that nothing else is at risk of catching fire. (The next illustration in step 7 shows where the little gear is soldered on the plate.)

Caution

Read and follow the instructions for whatever soldering compound you choose, and be extra certain that you have moved the denatured alcohol far away from the area where you are using the torch.

Step 7   Now we will add the large gear. Slot an 8-32 × 1 1/8-inch-long solid brass machine screw through the mounting hole from the *inside* of one of the side plates. Secure the screw in place with a brass washer and an 8-32 brass machine nut.

Step 8   Turn your attention to the other side plate. Position and mark a hole that will sit opposite the hole for the hand crank. After marking the spot for the hole, drill it out using an 11/64-inch drill bit. Then slot in an 8-32 × 3/4-inch brass machine screw. Secure it in place with a brass washer and 8-32 brass nut. Then add a second 8-32 nut atop the first.

We will use the 8-32 knurled nut to turn the machine screw into an "electric terminal." (More on that in Stage 8, later in the chapter.)

Step 9   Attach both side plates to the gun stock by slotting two 6-32 × 2-inch brass machine bolts through your 9/64-inch mounting holes in the plates. Use brass *finish washers* on the outside of both plates and secure firmly using 6-32 brass nuts. If the ends of your machine screws are too long, cut them to desired length or grind off the ends using your rotary tool.

Note

An alternative method of assembly would be to attach the brass side plates to the gun stock using an epoxy or high-strength adhesive of some sort. And it may well seem that this is an easier option. The problem, of course, is that once the piece is glued together it cannot be disassembled. At this stage, since the project is still evolving, we do not know whether we might still require access to the interior breech section of the gun stock. We want our project (like any real firearm) to be capable of being disassembled and reassembled. Remember the first principle of Steampunk mad science: Screwed is always better than glued.

Step 10   Return next to your first side plate—the one designed for the hand crank. Slide the large gear onto the post for the hand crank and secure in place with an 8-32 brass nut. Be certain to lock down this nut securely using a dab of Loctite adhesive or superglue so it does not vibrate loose through use and transportation.

Note

I recommend you use only Loctite or superglue, because they allow for the nut to still be loosened should you decide it necessary or desirable.

Step 11   Screw on your hand crank handle. Your project should now look similar to this:

Stage 6: Add the Sighting or Targeting Mechanism

Step 1   Now take apart all of the gun components so that you have once again your *raw* gun stock and barrel.

When the gun components were still assembled, I studied it carefully for inspiration. (This is what we refer to as an "evolving" project—a process that is applicable to a one-off art piece that does not necessarily translate well to production of multiples.) In studying the assemblage, I decided that I wanted to add a bracket and targeting mechanism to the back of the breech end of the gun in the area where the hammer would sit if it were a revolver (which of course it is not). But to add this bracket, I first need to flatten off the back end of the stock, just behind the breech. I proceed, therefore, to cut off the back curve (using my band saw with a wood cutting blade) in favor of a flatter shoulder and then sand it smooth.

Step 2   I selected a gear with a shaft to use as my sighting mechanism.

Note

Your selection might be different, but remember that this project is simply an example of the sort of gun that could be made depending on the nature and specific qualities of found objects. But for ease of explanation, let us continue to assume that you have components that are identical to those with which I am working.

Step 3   Select a strip of brass that is about 6 inches long, 3/4 inches wide, and 1/16-inch thick (available through craft and hobby shops) to use for the bracket.

Drill a hole about 1/4 inch from the end of the strip. This small hold needs to be sized to accommodate the end of the shaft that extends through the targeting gear. In this case, a 1/16-inch drill bit will suffice.

Step 4   Using a 3/8-inch drill bit, drill a hole through the strip to accommodate the end of the all thread rod that extends through the barrel (the lamp) and is capped by the breech finial.

Step 5   Now we want to design a shape for the bracket used to attach the site to the gun. Using a fine-point marker, outline a profile onto the brass strip.

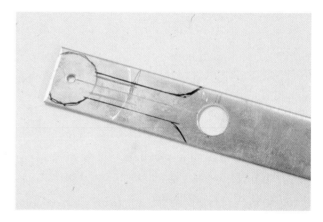

Then mark the position for a right-angle bend in the bracket. Then outline the bottom profile, which will sit against the gun stock (see the illustration in step 7 to get an idea of how this will look).

Step 6   At the bottom end of the bracket, mark a hole, then drill it out with a 11/64-drill bit to hold a #2 wood screw, which will help secure the bracket to the gun stock.

Step 7   Using a band saw or rotary tool, carefully cut out your bracket from the 3/4-inch brass strip. Then sand, file, or grind off all edges and corners. It should look something like this when you're finished:

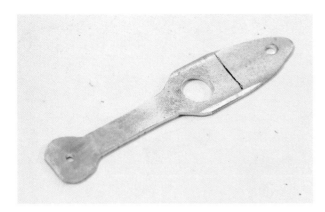

Step 8   Clamp the bracket tightly into a vise and heat the fold line carefully with your propane torch before using a hammer to bend it over to roughly 90 degrees. Continue this until the bracket base matches the profile of the gun stock while connected to the back of the breech.

Hint

Heating the brass bracket with the propane torch while bending it mitigates the possibility of it cracking or even snapping off. Bear in mind that bending and rebending brass will cause it to harden and stiffen, making it more brittle.

Step 9   Once the bracket has been bent to the proper angle (and cooled), proceed to solder the geared sighting mechanism to the bracket,

inserting the tail end of the shaft through the hole you made in the top of the bracket. It should look something like this:

Step 10   After the soldered assembly has cooled, place the sighting mechanism on the lamp all thread rod, add the breech finial, and then set it into place on the wooden gun stock using a #2 wood screw.

Stage 7: Add Finishes

Step 1   Remove the finial, the breech and barrel assembly, and the sighting mechanism from the stock. Lightly sand off any varnish on the gun stock to prepare the surface to accept a fresh coat of stain.

Step 2   Using a clean cloth, apply an even coat of red mahogany wood stain over the entire gun stock, being certain to wipe off any excess.

Step 3   Optionally, apply an even, thorough coat of Bombay mahogany stain and polyurethane (all in one) using a disposable paint brush. Or, after the step 2 stain has dried, apply a coat of polyurethane.

Step 4   While waiting for the stain and varnish to dry on the gun stock, use a sharp marker and accent the details on the cast brass barrel pieces. This will enhance the details and give it more character.

Step 5   Finally, you can "scorch" the muzzle end of the barrel using some flat black spray paint. You want the gun to look well used, rather than all bright and shiny.

Stage 8: "Electricification"

Yes, I know that is not a real word, but that is what we are doing to this project in this stage. We are "electricificating" it—creating the illusion of an electrified weapon, without actually having any electricity connected to it in any way, shape, or form. Write it down, use it in conversation, love it, and make it your own. Now back to building.

Step 1   Reassemble the breech, barrel, bracket, finial, and stock. You will then need three 12-inch long by 1/8-inch diameter brass rods, which you should be able to find at most hobby shops that sell materials for remote control cars or planes and train sets, and the like. Using pliers or by securing the rods in a bench vise, bend a 90-degree angle roughly 3/4 inch from one end of each rod.

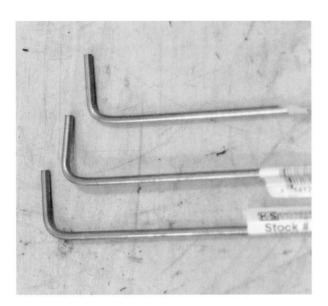

Step 2 Position one of the bent rods so that it can slide into the slot designed to accommodate the ramrod that was part of the original gun stock, just under the barrel. (If there is no such hole on your stock, drill an 1/8-inch hole in the corresponding location.) The bent end can be inserted into one of the "ventilation holes" around the muzzle of the barrel. (My lamp happens to have the perfect holes already in place; refer back to the illustration in step 5 of the previous stage, when we scorched the muzzle. If yours does not have these, drill an 1/8-inch hole into the base of the muzzle, aligned directly out from the ramrod hole in the stock.) Then mark the rod to a length so that 1–2 inches of the rod will extend into the ramrod slot. Do not install the rod just yet.

Step 3 Cut the rod to length using heavy duty wire cutters (such as linesman pliers or fencing pliers), or a band saw or hacksaw.

Step 4 Using a pair of wire cutters, cut a piece of copper wire (easily obtained from anywhere that sells electrical supplies; I use stripped housing wire) roughly three times the length of the brass rod, and then coil the wire around the rod. Do not wind it too tightly. Use your wire cutters to cut off the excess length of copper wire. When you are finished it should look like the image at the bottom of this page.

Step 5 Slide the straight end of the rod and coil into the ramrod hole and slot the angled end of the rod and coil into the hole at the base of the muzzle, as shown at the top of the next page. Superglue or solder in place as necessary.

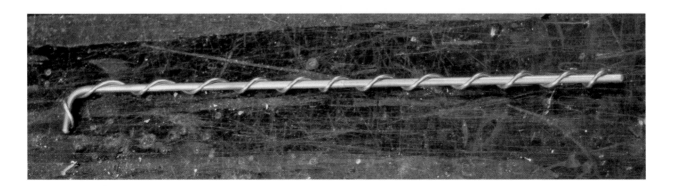

Step 6   Now take the second brass rod and place the angled end into a hole at the base of the muzzle about one third of the way around from the first rod. (Again, if your muzzle does not have these holes already, drill one, as described in step 2.) Run the long end straight back over the breech chamber. Place a mark on the breech chamber 1 inch forward of the end of the rod. Remove the rod. Using an 1/8-inch drill bit, drill a hole on the mark you just made. Repeat this operation for the remaining rod, about one third of the way around from the bottom rod but on the other side of the barrel.

Caution

Be certain to use your spring-loaded center punch to dimple the brass before attempting to drill it out, because you're working with a smooth compound curve, and the drill bit could easily slide off or "walk" on you. And it would be a terrible shame to destroy your project at this stage. You may find it easier to remove the breech from the stock to drill these holes.

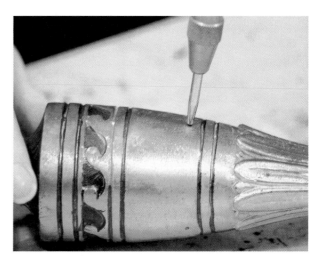

Step 7   If you disassembled the device to drill the holes for the rods, please reassemble it now. We need to make a second bend at the remaining straight end of both rods after we have determined the desired length. Do this by replacing the angled end of a rod into the same hole near the muzzle you had it in before with the long end passing over

the corresponding hole on the breech. Mark the rod directly over the breech hole. Remove the rod. Bend the end of the rod at your mark in the same direction as the other bent end. If you make this bend a bit tighter than 90 degrees (say, about 100 degrees), the rods will sort of "clip" into place and you might get away without having to solder or glue them into place. Check the fit by placing the first end in the appropriate hole near the muzzle and the last end into the hole in the breech chamber.

Step 8   Coil the copper wire around the two brass rods. Then clip, solder, or otherwise secure the two brass rods into position, as shown immediately below. (I removed the breech/barrel assembly for the picture. You do not need to do so.)

Hint

You can use a pair of pliers or cutters to sort of kink a small notch into the rod or coil wire to help

secure it in place on the gun barrel. To do this, pinch the rod or coil between the jaws of a cutter and then rock the piece back and forth gently. This should score the coil without cutting through it and create a little shoulder of sorts.

Step 9   If you have yet again taken the blasted thing apart, assemble it all back together again and make sure it looks pretty close to what you see in the second image at the bottom of this page.

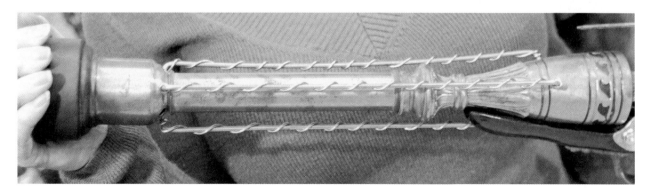

Step 10   Because you are of course using the same parts that I am, you will see three holes in the blunderbuss-shaped shroud at the muzzle end of the barrel. They were originally for screws to hold the lamp shade to the lamp. We want to fill each of these holes using a 6-32 × 1/2-inch-long brass machine bolt, a washer, and a 6-32 brass nut. Insert the bolts and washers into the holes, and finger-tighten the nuts onto the machine bolts. (It's shown here without the scorched effect or other parts to show greater detail.)

Step 11   Add a second lamp finial to the all thread rod inside the blunderbuss muzzle end of the shroud. I recommend something pointed, or even a crystal sphere for this purpose, because it should look like some sort of focal point for the lighting discharge.

Step 12   I had a spare finial lying around with a ring on the end of it. This sparked an idea.

After attaching the finial, I cut three lengths of braided copper wire to roughly 12-inch lengths. I attached the end of the wires to the screws (under the washers) on the inside of the muzzle shroud.

I then proceeded to wind the wires around and through the finial loop as shown here:

Step 13   Now we'll add some "electrical coils" to the assembly. (We'll use similar coils in some of the other projects of this book.) I begin by selecting a Phillips head screwdriver. This will serve as a mandrel around which we will coil the wire. (Phillips head screwdrivers are good to use because slot-headed screwdrivers flair at the tip,

which makes it harder to slide off the coil.) You can use any sort of rod that you wish, but bear in mind that its outer diameter will dictate the inner diameter of the coil, so choose wisely.

Step 14   As a rule, 3 inches of wire equates to about 1 inch of tight coil. I want to make tight coils that are about 1 inch long when finished, with about 1 inch of straight wire at either end. In this case, each length of copper wire for these coils should be about 5 or 6 inches long. (You might add an inch or two for good measure.) Tightly coil the wire around the mandrel, as shown, and then slide the coil off.

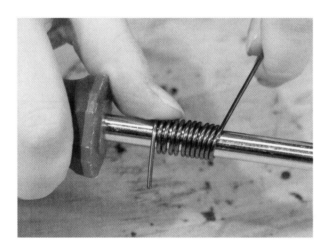

Step 15   After the coils are the size you want, trim off the straight ends of the coil wire so that each is about 1 inch long. Then bend the ends around 180 degrees.

Step 16   Secure one end of a tight coil under the nut and washer of one of the terminal posts on the muzzle of the barrel. Lock it down using a 6-32 knurled nut.

Step 17   Stretch out the coil like a spring, and hook the end into the empty ventilation hole between two of the three rods.

Step 18   Repeat steps 16 and 17 for each of the coils connected to your muzzle end terminal posts. When you are finished, your project should look like the image at the top of the next page.

Step 19   For further "electricification," let's add three additional coils on the plate attached to the side of the breech section (see the image in the next column). Remember this bit that we left empty except for a single knurled nut (Stage 5, step 8)?

Make three more coils by repeating steps 14 and 15, and secure all three of them under the knurled nut on the side of the gunstock. See image at the bottom of this page.

Step 20   Stretch out the coils as before and wind the loose ends around the breech end of the rods that we attached earlier (see top of next page).

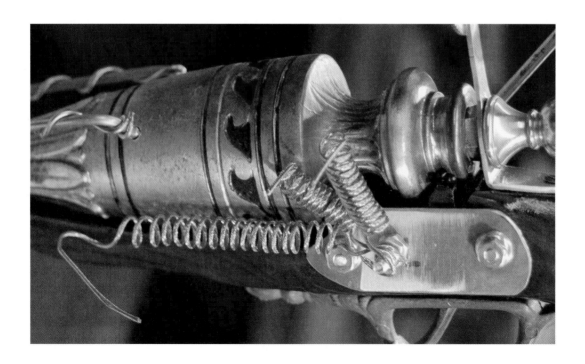

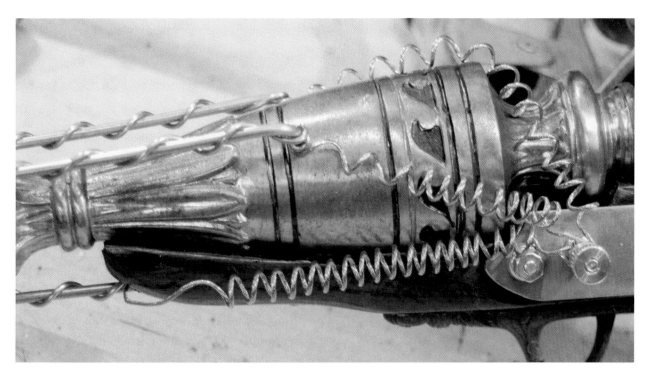

Stage 9: Add Ornamentation and Adornment

All of our "functional" mechanisms are now complete. One of the beautiful aspects of Steampunk is that its properly designed accoutrements are simultaneously practical and impractical. The design elements appear as though they would actually work, but not necessarily in the simplest or most practical way (such as the exposed "electrical" coils we have just added).

To make our High Voltage Electro-Static Hand Cannon look even more timelessly Victorian in design, we need to add some flourishes of decoration and adornment. The late 19th century was rife with ornament. No bare surface would be left without some sort of decoration, pattern, or texture, which was simply the aesthetic of the era. To that end, I have selected a few little metal ornaments from my junk box, a jewelry supply store, and the local scrapbooking supply shop.

Step 1 Choose your adornments. I have elected to use a pair of rectangular brass ornaments for the pistol grip and a brass lion mask (commonly produced, or reproduced, for furniture restoration) on the butt of the stock. Use a drill fitted with a 1/16-inch drill bit to create two screw holes for each piece of ornamentation (so that the ornament does not spin, rotate, or change position).

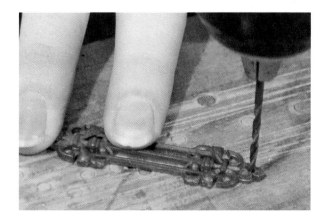

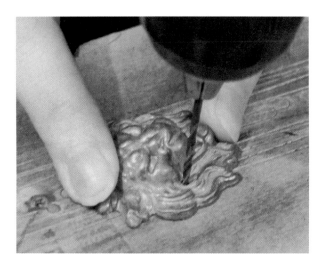

Step 2 Once the lion mask (or the ornament of choice) has been drilled out, use two sets of pliers to bend the ornament slightly. This will help it conform to the profile of the butt of the pistol grip.

Make holes for the screws in the gun stock using your awl or by drilling small holes with a $\frac{1}{16}$-inch drill bit.

Caution

Be careful not to bend and rebend the metal too many times, lest it become brittle and break. Proceed slowly and carefully as you bend.

Step 4 Fix the ornaments into place using small #2 brass wood screws.

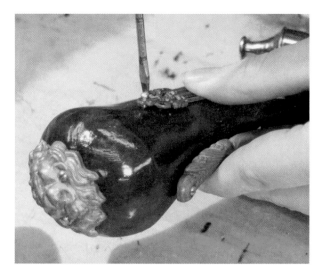

Step 3 Position the ornaments onto the gun stock one at a time and use an awl to mark the location of the screw holes in the stock.

Of course you could keep adding more and more ornament and decoration and complication to the project, but as far as this particular contraption is concerned, I think I have nearly reached a stopping point. Just a couple little final touches and then the project can be put to bed.

Aesthetic Details and Authenticity

Although it is not likely to matter to most people, I make a point when working on Steampunk projects to use slot-headed screws for any visible hardware whenever possible, instead of Phillips head screws. Henry Phillips did not invent the Phillips head screw until the 1930s. Therefore any Steampunk design that pretends to be Victorian (or Victorian-ish) would not have Phillips head screws. It is a matter of aesthetic detail that few might consider necessary, but in the words of the great English playwright, Tom Stoppard, "An audience knows what to expect and that is all they are prepared to believe in." That is to say, someone looking at it might not know quite why it looks wrong, but the person is likely to register that something is off, or not quite right. So these little details, such as using the correct sort of screw head, can indeed make a difference.

Stage 10: Finish It Up

To finish it off, add another light layer of flat black spray paint over the "scorched" end of the muzzle.

The idea is to add the scorched effect to the brass terminal posts and coiled wire. It was important that you already "scorched" the muzzle earlier, because if you had waited until this stage to do that, you would likely get odd shadows and occlusions in the paint from the coils and posts.

Finally, give the whole project a lightweight coating of clear polyurethane to help minimize chipping, damage, and oxidation.

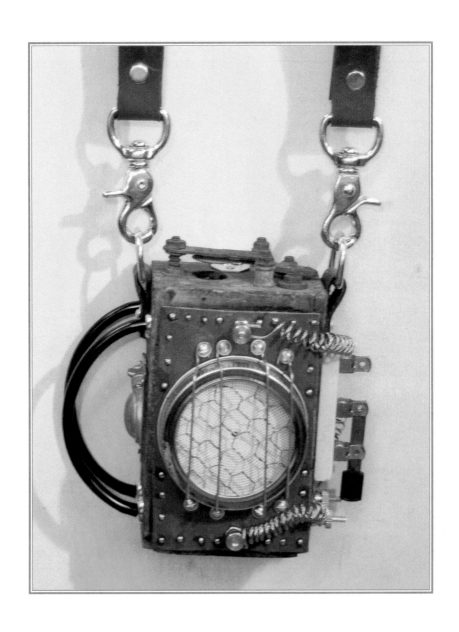

Chapter 9

Tesla-pod Chrono-Static Insulating Field Generator (or, The Mobile Device Enclosure)

"I can't believe we are trapped here. We won't even be born for another 120 years," said Major Smythe with a hint of panic in his voice.

"We must get out of this room. According to the book, this building will be destroyed in an unexplained explosion this afternoon," said Professor Grimmelore, ignoring the Major's look of distress.

"Leave this room? We can't do that!" interjected Ms. Grayson. "We have traveled back in time. Any interaction with this time's denizens could change everything. We are doomed! There is no..."

"Get hold of yourself, madam," commanded the Professor, interrupting Ms. Grayson's slip toward madness that comes with dealing with chrono-dynamics. "We will be fine. We all have our Tesla-pods, right? They all have the blue light on, correct?" Everyone looked and nodded.

"Good! As long as your T-pods are functioning, nothing you do here will affect our future, remember?" concluded Dr. Grimmelore, inspiring a collective sigh of relief. "Alright, Feathers, open the door and take point. We need to get to the street as quietly as possible."

"Yesh Sa'!" said Featherstone, far too loudly for everyone in the room. Grimmelore was pretty sure Featherstone wasn't drunk, but it was sometimes difficult to tell with him.

Lord Featherstone's mechanical arm twisted the lock from the door as though he were simply plucking a flower from its stem. The door flew open, and a man with a sword yelled at the group in what was most likely guttural archaic French; he swung his sword at Featherstone.

"Shut your bloody pie hole, froggy!" yelled Feathers, as his natural hand collided with the Frenchman's natural and evidently rather fragile jaw. The Frenchman hit the wall on the other side of the hallway and slid to the floor like a sack of rocks.

"Gods, man! Could you make more noise? Please try and have a modicum of discretion." Grimmelore's reprimand did nothing to erase the self-satisfied look on Featherstone's face. "Let us get out of this place before things get worse."

They made their way cautiously down the hall and could see what appeared to be an open door to the street.

"So, Doctor...um...what happens if the little blue light goes out?" inquired Featherstone almost sheepishly.

The doctor turned toward Featherstone with a sense of absolute dread and noticed a slash like that of a sword across the front of Featherstone's T-pod. The lit display was blinking erratically, as small sparks cracked along its coils.

"Oh, bugger..."

–From *Featherstone and Other Unnatural Disasters:*
My Travels with a Mad Man, by Professor Adrian Grimmelore

*W*hat is a Tesla-pod? It is a chrono-static insulating field generator, of course—an absolute necessity for any time-traveler. This device helps prevent those pesky popping-out-of-existence issues that can always occur when encountering an all-too-frequent time paradox.

Suppose, for example, that during your travels back in time, you inadvertently prevent your grandfather from meeting your grandmother, thereby negating the existence of your own parent and, consequently, yourself. A chrono-static insulator will help protect you from such paradoxical catastrophes and temporal anomalies. Or think of it this way: Imagine a goldfish being scooped from its tank and placed in a bowl of water. Now imagine that all the water in its original tank is replaced with fresh water, and then the goldfish is reintroduced to its original home, safe and sound. The smaller fishbowl created a temporary protective environment around the fish while his normal environment was not available. The Tesla-pod chrono-static insulating field generator creates a sort of bubble around you, similar to the smaller fishbowl. When you return to your present time, therefore, you will be unaffected by any changes you might have made to the past. The same cannot be said about your environment, however: You might return to a present time in which your family members never existed and it is raining marmalade or some other unpredictable consequence, but you will, at least, be the same as you were before you began your time travels (assuming, of course, that you stayed within the chrono-stasis field generated by your Tesla-pod at all times during your travels.)

Project Description

In this chapter, we will be addressing the problem of what to do with that annoying piece of portable technology you are carrying on your person at all times. I know you could put it in your pocket, but then there is that unsightly bulge, and most skirts and dresses do not even have pockets. In the finest Steampunk tradition, we will construct something terribly complex to solve a simple problem. Here we will make a cool case for your phone or media player.

Note

Of course, mobile phones are constantly changing and being redesigned. Some models will fit in the case, while others will not. Some will fit with their protective cases, and others will not. So what you choose to put inside of your T-pod and how you intend to use it is entirely up to you.

For this project, we use a Model T Ford ignition coil as the T-pod case. Other sorts of wooden boxes could be used (or made) to the same effect, of course, and these would have different dimensions (internally and externally) and could be used to house different sizes of modern electrical conveniences (or necessities). But a Model T coil looks so cool with all those bits on the top. We have, for example, on at least one occasion, redesigned one of these T-pod units to house an insulin pump for a diabetic client.

Note

Read through the entire chapter before beginning this project, because what we have presented here is just one possible variation of T-pod design. As your design may vary from ours, you might require different tools and materials to those listed here.

What You'll Need

To complete the project as described in this chapter, you will need to get your hands on the following materials and tools.

Materials

- ❧ Model T Ford ignition coil
- ❧ Pocket plasma clip-on light
- ❧ 5–6 oz. leather piece (at least 5 × 8 inches) any color
- ❧ 1 standard size Mason jar lid band (2.8 inches)
- ❧ Light metal screening
- ❧ Brass rod or 14-gauge wire (I use a 3/64-inch brass rod)
- ❧ 6, 3/4-inch, 6-32 brass bolts
- ❧ 8, #6 brass washers
- ❧ 8, 6-32 brass nuts
- ❧ 4, 6-32 knurled nuts
- ❧ 8, 1/2 inch, 4-40 brass bolts
- ❧ 8, #4 brass washers
- ❧ 8, 4-40 brass nuts
- ❧ 3/4-inch × 2 3/4-inch brass hasp with hook
- ❧ Small brass rivets (optional)
- ❧ Single blade knife switch
- ❧ 2, 1-inch 6-32 brass bolts
- ❧ 10 inches of 1/4-inch black plastic tubing
- ❧ 4, #10 wood screws
- ❧ 4, #10 brass finish washers
- ❧ Handful of small brass brad nails
- ❧ 3 feet of braided copper wire (or other wire)
- ❧ Strip of leather, 3/4-inch wide × 30 inches long (optional)
- ❧ 2 brass "D" rings (or suitable substitute, optional)
- ❧ 2 brass snap hooks (optional)
- ❧ Double-capped rivets (optional)

Tools

- ❧ Safety glasses (of course)
- ❧ Small brad puller
- ❧ Nitrile or latex gloves
- ❧ Small brad hammer
- ❧ Small wire cutters
- ❧ Scraper
- ❧ Medium-grit sandpaper
- ❧ Sharp marker
- ❧ 1/4-inch drill bit
- ❧ Power drill
- ❧ Quick clamp
- ❧ Coping saw (or reciprocating power saw)
- ❧ Leather scissors
- ❧ 1/4-inch disposable paintbrush
- ❧ Contact cement
- ❧ Utility scissors
- ❧ Needle-nose pliers
- ❧ 1/8-inch drill bit
- ❧ 5/32-inch drill bit
- ❧ Superglue or Loctite adhesive
- ❧ White or wood glue
- ❧ 1/16-inch drill bit
- ❧ 13/64-inch drill bit
- ❧ Diagonal cutters
- ❧ Large screwdriver
- ❧ Center punch

Alternative Tools

- ❧ Rotary tool with cut-off wheel
- ❧ Rotary tool with sanding disc

Stage 1: Take Apart and Clean Out the Case

Here is the Model T ignition coil, in its natural state: dusty and probably broken and such. I acquired this one at my local flea market. You can also find them on eBay or from other Internet sources. We are going to tear apart the case and clean out all the nasty stuff inside.

Step 1   We'll start by removing one of the large wooden panels. Notice the two small nails at the bottom of one panel; you want to remove these using a "cat's paw" tool—a tiny little pry bar, also called a "brad puller." (I have named mine Janet—as in *Rocky Horror*.) Slip the forked tines of the tool under the edge of the panel and pry it up just a tiny bit and then let it back down. When the panel slips back down, the little nails should stay up a bit so you can grab them and pry them out with the brad puller. Remove both nails, of course, and remove the panel. Throughout the rest of this project, this panel will be called the *back door*.

Alternative Tool: Rotary Tool with Cut-Off Wheel   So you tried the cat's paw and the little nail broke off, slid through the wood, wouldn't come out, or whatever. Fear not, for there is another way. Grab your rotary tool and attach one of those handy cut-off wheels. I'm sure you are already wearing your safety glasses. You can cut off the nails by cutting into the slot under the edge of the panel and through the nail. See how easy that was?

Step 2   Slide the panel out, and you'll see that the case is filled with dried black goo. I have no idea what this is, but it *is* the inside of a battery, so it is probably pretty nasty stuff. *Be sure, therefore, to put on some goggles and nitrile or latex gloves before you handle this stuff.* You need to remove the goo and anything else in there by taking the box completely apart.

Place the case flat on the table with the open side up and the bottom nearest you. Place the tips of your brad puller down along the inside of the side of the box near the finger joints in the lower left corner. Gently tap the handle of the puller with a small brad hammer. The finger joint will begin to separate. Use the puller to pry it apart slowly, and the side should open up as shown here:

Step 3 Pry off the bottom from the open side toward the other jointed corner. As you do this, you might notice a wire soldered to a metal spot on the inside of the bottom piece. Using a small pair of wire cutters, cut it off close to the bottom piece.

Step 4 Work your way around the box removing the sides and the top, cutting the wires as needed.

Hint

The case fits back together only one way, so keep track of the box pieces. As you take it apart, draw a little arrow on the inside of each piece to indicate the top of the coil—the part with the metal terminals on it.

Step 5 If you are lucky, the black goo will be one big lump. Pry off the back panel of the box (which, from now on, will be called the *front* of the case) and remove the goo. Yes, I know there are some interesting looking bits sticking out, and I am a pack rat like you, but do you know what that stuff is? Neither do I. So wrap it up and take it to a hazardous waste facility; do not throw it in the trash.

Caution

The internal parts of each coil were sealed against moisture by this black insulating, tarlike compound, called "Ford Hydrolene." Nobody knows exactly what Ford put in this stuff, so its toxicity is not known. Please do not throw this stuff in the trash. Instead, take it to a hazardous waste facility (most counties or townships have them) and ask them to dispose of it properly.

Step 6 Bits of stuff will probably be stuck to the inside of the case. Use a scraper to remove them and sand the inside and outside of the case with a medium-grit paper to clean it up a bit. If your case parts look something like the following image, you are on the right track.

Stage 2: Cut Out the Display Portal

I love to use pocket plasma clip-on lights in my Steampunk projects. They can be set to be turned on constantly or to be sound activated. I usually buy them from online vendors.

Step 1 Place the front panel on the table with the arrow pointing away from you. Flip the panel so that the arrow is on the underside. You are now facing at the outside of the front of the coil case. Center the clip-on light on the front panel, with the display (the white part) facing up. Then slide the light up (away from you) about 1/2 inch. Using a marker, trace around the light's battery compartment—the black box on the back.

Step 2 Now to cut out that part you traced. Insert a 1/4-inch bit into your power drill. Drill a hole in each corner of the square you traced.

Hint

Place a piece of wood under the front panel so you do not destroy your table when you drill your holes.

Step 3 Clamp the front panel to the corner of your work table, with one of the borders of the traced rectangle just barely off the edge of the table.

Step 4 Remove the blade from your coping saw. Run the blade through one of the holes, and reinstall the blade back in the frame.

Hint

Make certain the teeth of the blade are pointing in the right direction when you place it back in the frame so you can make the cut.

Step 5 Saw along the first line (the one hanging over the table). Unclamp the piece and rotate it so another line is hanging over. Cut all four lines of the rectangle until the piece is free. See what I mean?

Stage 3: Make the Leather Front Covering

Step 1   Place the leather on the table with the top side facing down. Place the front panel of the case, inside up, on top of the leather. Trace the outside edges of the panel onto the leather. Trace the inside of the cutout onto the leather. Remove the case panel, and then draw an arrow on the piece of leather (between the cutout and the sides) that corresponds with the arrow on the front panel.

Step 2   Using leather scissors, cut the leather along the outside lines.

Step 3   Cut out the hole from the piece, but cut just barely inside the lines. This is where you will be mounting the light.

Step 4   On the outside of the front panel and the back side of the leather, use a 1/4-inch disposable paintbrush to apply a light coating of contact cement, 1/4 inch wide, around the edges of the hole. Let this dry for about 15 minutes. When the glue on both pieces feels tacky but not wet, put the two glued sides together. Make sure the outer edges match up and the arrows you drew on both pieces are facing the same direction.

Stage 4: Add the Bezel

Some of you might be wondering, "What is a bezel?" The rest are probably looking it up online. A bezel is the rim that holds the transparent covering of a watch or headlight, for example. On an old clock, the brass part holding the glass over the face is the bezel. Sometimes I use a bezel from an old clock on a project like this. If you don't happen to have 30 or 40 old clocks sitting around waiting to be pillaged, no problem. I have a great substitute. A standard size (2.8 inches) Mason jar band (from the lid) fits your pocket plasma light almost perfectly.

Step 1   Slide the battery compartment of the plasma light through the hole in the panel from the outside, with the switch toward the bottom. Then set this to one side for a bit.

Step 2   Lay some wire screening (mesh) on the table. Place the jar band on top of the mesh and trace around its outer edge with your marker.

Hint

You can use all different types of mesh for this. Look around at a craft store and see what they have, or use old window screen. Even better, some hardware stores sell brass screen. Old-style fishing supply stores might have odd screens for fishing traps and bait boxes.

Step 3   Use a pair of heavy utility scissors to cut out the mesh. Insert the mesh into the Mason jar band, as shown on the top of the next page.

Step 4 The cool thing about using this jar band is it will actually screw onto the front of the light. So that is what you will do now. It should look something like this:

Stage 5: Install Mounting Hardware

This is the stuff that will keep the light from moving around. This is also one of the most challenging parts to describe. As we will not be using exact measurements, you are going to have to do whatever looks good for your project. I'm going for a protective wire cage look, with four bars across the front of the light.

I'll use 3/64-inch brass rod for some of this. Brass rod can be obtained from hobby shops that sell remote control cars and planes. But, equally, you could use 14-gauge copper housing wire or heavy brass wire from the craft store as well.

Note

Hmm. This will be difficult to explain, but I will do my best, if you promise to do your best to follow along. Agreed? Okay, then, let's proceed.

Step 1 Cut four wires (or rods) for your cage, each the same length; mine are 5 inches. (If you are using something other than a Mason jar lid, this length might vary as should become apparent in the following steps.)

Hint

Rather than measuring each piece individually, try marking your measurement on a piece of paper; then extend the wire on the paper and cut each wire to size.

Note

You would be unwise to dispose of any remaining wire or rod. You might need to use them later in this project or in other projects. As a general rule, keep everything—scraps can be very useful.

Step 2 Using a pair of small needle-nose pliers, bend one end of each rod, as shown in the illustration. These "loops" will be screwed to the front of the case, so make the bends a size that will fit the screws you'll use for this purpose.

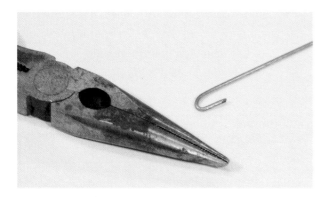

Step 3 Bend the rod about 90 degrees just before the first bend. (See the image in step 5.)

Step 4 Remove the bezel (with screen) from the front of the light (assuming you have already put it on there).

Step 5 Place the bezel on a smooth, flat surface, and stand the rod next to the bezel, as shown. Mark the height for another bend to go over the bezel.

Step 6 Bend the wire in a sort of Z shape, so it looks like this:

Hint

Bending the wire slightly shorter than the height of the bezel will help tighten and secure the entire apparatus.

Step 7 Repeat step 6 with a few more rods (I've used four in all). Then assemble the component pieces and position your rods (or wires) to form four chords (if memory of my high school geometry is correct).

You can, of course, choose any type of design here. This is, after all, *your* project. Whichever configuration you choose, the principles and process of attachment will be similar.

Note

You can achieve different effects by using different types of metal wire or rods.

Step 8   Mark the position of the loops for your screw holes on the leather mounted on the front of the case.

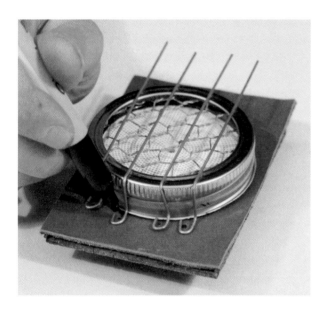

Step 9   Using the outer rim of the bezel as your guide, mark the other end of the rods, where you'll make the next bends.

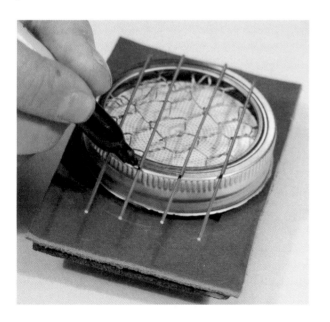

Step 10   At the mark, bend the rods down 90 degrees.

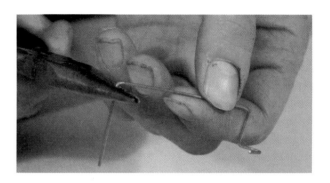

Hint

Do yourself a favor, and keep track of which rod goes where, because they are likely to be of different lengths—that is, they are not entirely interchangeable.

Step 11   Mark the bezel height on the rods, below the last bend, and bend them all in the same Z shape and loop you used for other ends of the rods. Note again that making this height slightly shorter (and I do mean *slightly*) than the bezel height will help secure and tighten down the whole assembly.

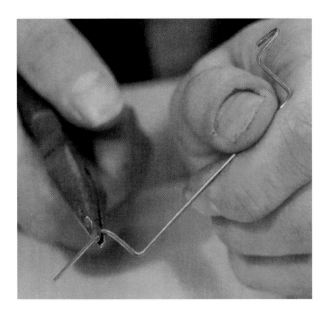

Step 12   Bend the last loop as you did in step 2.

Because the pieces you are bending are shorter and you have less leverage, you might find it useful to use your work surface as an aid in bending the rods around your pliers.

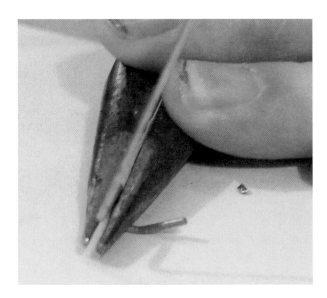

Step 13   Cut off any excess wire. You should now have four rods that look like this:

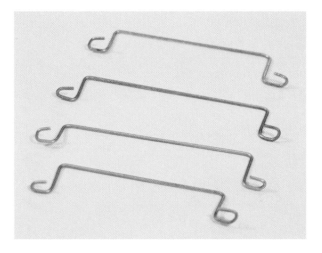

Step 14   Holding each rod on the bezel in turn, match the first loop with the marks you made in the leather in step 8. Then mark the position of the other end loop indicating where to drill the screw holes through the front.

Note

Yes, you could measure all of this out using calipers and micrometers and such and achieve factory precision, but bear in mind that this is an unholy alliance between art and science, and as such, improvisation is not only welcome, but encouraged.

Step 15   Disassemble everything and set the parts aside. Having already marked the position of your screw holes on the front of the case, install a 1/8-inch drill bit into your power drill and drill the holes through the leather and wood of the case front, as shown in the next image.

Note

On occasion, these antique wooden panels are brittle and prone to splitting. If this happens (as it sometimes does to me), simply use a gap-filling superglue or epoxy to stabilize it. Because this particular panel is hidden by the leather, any splits in the wood will never be noticed anyway.

Stage 6: Add More Ornament: The Terminals

If you are planning to incorporate more ornamentation on the front in addition to the light/bezel/cage assembly, now is the time to mark their positions and attach them in whatever way seems most practical. I have chosen to add two electrical connection terminals to give the whole unit an even more functional appearance.

Step 1   To make these terminals, we use a 5/32-inch drill bit to bore two holes at the center top and bottom of the front panel. Insert a #6 brass washer over a 3/4-inch, 6-32 brass bolt and pass the bolt through one of the holes. Secure it in place with a #6 brass washer and 6-32 brass nut.

Note
The washers should be on the inside face to help ensure that the wood does not crack or split when you tighten down the nut. Be certain to lock each nut using a dab of superglue or Loctite adhesive (available at most hardware stores).

To complete the other "terminal," add a second brass washer and then a 6-32 knurled nut.

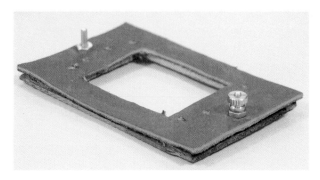

Step 2   Slot your light unit back into the front of the case, and reposition the bezel. It is now time to attach the wire cage. Position the rods and slot 1/2-inch 4-40 brass bolts through the drilled holes and wire loops. Secure them with #4 brass washers on the outside and 4-40 brass nuts on the inside.

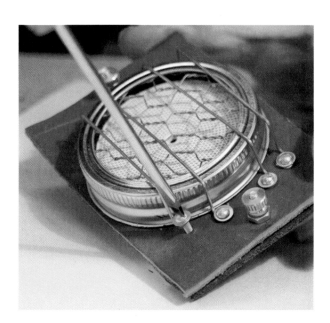

Note
Be certain to lock down these nuts with a dab of superglue or Loctite so that they do not loosen up through vibration of wear and use. (Alternatively, lock washers might well be suitable to the task if you have some on hand.)

Stage 7: Reassemble the Case

Step 1 Using a small amount of standard white glue (or wood glue) between each of the fingers of the case's finger joints, reassemble the top and sides of your case, referencing the arrows you marked on the inside.

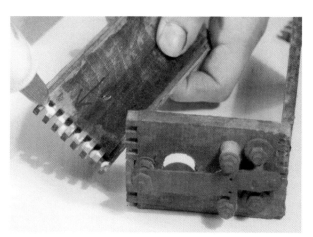

The front of the case is now complete. If your project looks like the following, congratulations. Let's proceed.

Note

Remember that these antique battery casings were each individually handmade, and not only are no two the same, but none of the pieces are interchangeable (which is rather odd considering that they were all made for Ford, America's greatest proponent of interchangeable parts). That is to say, they will go back together only if they are assembled in the same configuration used before you took the case apart. Hopefully, you added the arrows as you were taking apart the case. If not, you will have to solve this little three-dimensional puzzle on your own.

Step 2 After you have assembled the sides, run a thin line of glue in the front channel and slot the front into place, as shown next.

Step 3 Further secure the faceplate into place by attaching the bottom of the box using wood glue in the corresponding finger joints. Use a damp cloth to wipe off any excess glue.

Caution

Take your time with this step, because it can be tricky and is fundamentally crucial to the strength, structure, and stability of the finished case.

Step 4 Now comes the most challenging step of all: stop. Put down the case and walk away. The glue needs to dry and set for several hours—preferably overnight. If you have suitable clamps at

hand, they are a useful addition in this drying process to prevent the box twisting or the finger joints separating while the glue is drying.

Stage 8: Add a Hasp to the Back Door Panel

Step 1 Clean the channel backing the back door panel to remove dirt, debris, or glue and slide the panel into place.

Step 2 Position a 3/4-inch × 2 3/4-inch hasp with hook into place on the top-middle of the back door panel and mark the position of the screw holes.

Note

Virtually all packaged hasp and hooks come with mounting hardware. Depending on the length of the screws that came with your hasp, and depending on the thickness of the wood on the top panel of your box (the one with the antique terminals), you might need to find shorter screws, or alternatively cut them down or grind/file them off.

If, as sometimes happens, the two hindmost screws align with the interior channel, you have a few options:

⟋ You could fill those problematic holes on the hasp with small rivets (as discussed elsewhere in this book), relying on one screw to secure the hasp, like so:

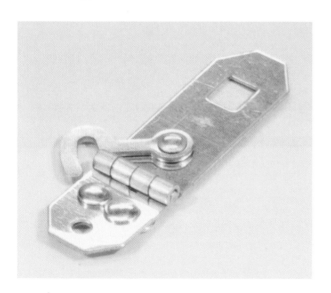

⟋ Drill new holes into your hasp hinge to reposition the placement of the screws.

⟋ Use superglue to affix the hasp instead of screws. (Not recommended: this is the most likely method of attachment to fail and break off. As you know, one of my primary axioms is "Screwed is always better than glued.")

⟋ Attempt all of the above.

Just make sure that the screws you use to attach the hasp and hook (which will hold the back door panel in place) do not interfere with the door panel sliding into the case.

Step 3   Using a 1/16-inch drill bit, drill pilot holes for the screws. Use the screws that came with the hasp, or other screws, to attach the hasp to the top of the case. Slide the back door panel into place.

Step 4   On the panel, position and mark the placement of the loop for the hasp hook.

Step 5   Remove the panel from the unit and drill out the holes using a 1/8-inch drill bit.

Step 6   You now have several options by which you can attach the hasp loop to the back door panel:

⟋ You could enlarge the holes in the loop hardware to accommodate small rivets (which is what I have chosen to do for this example).

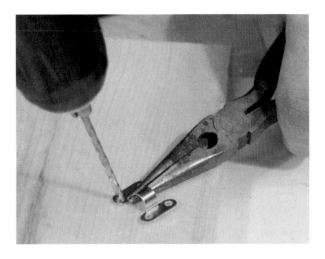

∽ You could use short screws (which would not protrude into the interior of the box).

∽ You could use #2 bolts and nuts to secure it in place, providing they are extremely low profile.

∽ You could glue the loop hardware into place (not recommended).

∽ Or you could simply use a fine gauge floral wire to wire the loop into place through the pilot holes you have drilled.

Caution

If you decide to use a drill to enlarge the holes, make sure that you grip the loop firmly with pliers or a vise before attempting to drill out the holes in the brass. Failure to secure the loop could result in the piece binding on the drill bit and becoming a finger-shredding implement of doom.

Whatever attachment method you use, make sure that the inner face of the back door plate is smooth and flush so that it does not interfere with the door sliding in or out of the channel. If the plate interferes with the door, you will have to notch out a passage for your mounting hardware, as discussed in step 7.

Step 7   Because I'm using small rivets to secure the hasp loop onto the back door panel, I have an extra step to perform. I found that the rivets catch slightly on the bottom panel as I attempt to slide it in or out of the channel. To compensate for this, I slide the panel partially into place and mark the location of the interference, so that I can create a shallow notch that will allow the door to slide past unhindered.

To cut out this notch, I use a standard electric rotary tool with a sanding disc attached to it, though the skilled reader might choose to use a small sharp knife or chisel to cut out the notch.

Step 8 Now we can go ahead and assemble all the parts. I suggest that you take a moment to step back and marvel at your own awesomeness.

You are now ready to add further adornment.

Stage 9: Add Adornment and Ornamentation

It probably goes without saying that the way you choose to ornament and fancify your T-pod is a matter of individual taste as well as what materials you have available. In this case, I will add a single blade knife switch (available from most electrical suppliers or online) as well as two additional "electrical contact terminals," along with some mysterious tubing and some wire coils.

Step 1 We'll locate the knife switch on the side of the case not occupied by the battery's contact points. Center the knife switch low enough so that it does not interfere with the hanging hardware (should you decide to add it, see Stage 11) or terminals (see step 4, coming up). Use a nail or marker to mark the placement of screw holes where the knife switch will be attached to the side panel of the case.

Step 2 Remove the back door panel and, using a 1/8-inch drill bit, drill out the knife switch mounting holes on the side panel.

Step 3 Attach the knife switch using 3/4-inch, 6-32 brass bolts with 6-32 brass nuts glue-locked into place.

Note

Some commercially available knife switches (such as the one I am using for this example) have a ceramic base as their insulator. Be careful not to overtighten these bolts lest you risk cracking the ceramic base.

Step 4 Now decide where to place two "electrical contact terminals" below the knife switch. Drill out the holes and repeat Stage 6, step 1 (when we added "electrical connection terminals" to the front panel). The only difference between these side-mounted terminals and the earlier set is that we now use 1-inch brass bolts instead of 3/4-inch brass bolts. The side panel, with the knife switch and terminals in place, should look similar to this:

Step 5 Now, turning our attention to the last side panel (the side with the battery contact points), we will add two, 5-inch lengths of 1/4-inch black plastic tubing (available from most reputable hardware centers).

Step 6 Mark the positions where the ends of the tubing will be attached to the panel. Mark four holes roughly 3/4 inch from each end of the case and equidistant from the front and back edges. Then carefully drill out the holes using a 13/64-inch drill bit.

Step 7 Slot four #10 woodscrews through these holes, from the inside out. Then slip on four #10 brass finish washers and work the plastic tubing over the exposed screw threads.

Step 8 Now we'll add a flush-mount drawer pull that I found at a local hardware superstore. Install it with its outer face on the inside, because this way it looks like a protected power port or data port. And we really don't have any use here for a drawer pull.

Step 9 Mark the position of the screw holes. (If the tubing is in the way, you can take it off for the time being.)

Using a ¹⁄₁₆-inch drill bit, drill out the screw holes.

Hint

Depending on the size of the screws you are using, be certain to select a drill bit that is marginally smaller than the screw, so that you leave enough material for the screws to bite into.

Step 10   Attach the pull using small wood screws. Your project should now look something like this:

Note

Use screws that don't protrude too far into the interior of the case. If they do protrude into the interior of the case, be certain to grind, file, or cut them off lest they scratch up any electronic devices you might hide within the Tesla-pod.

Stage 10: Add Rivets Around the Edges

Let's add a decorative rivet effect around the edge of the leather fascia using small brass brads.

Step 1   About 1⁄8 inch in from each corner, mark corner points, and then work your way around the perimeter, making marks at roughly 1/4-inch intervals.

Step 2   Using a 1/16-inch drill bit, carefully drill small pilot holes for the brads.

Step 3   Because the brads are slightly longer than the thickness of the front panel and would poke through into the interior, cut off the points of the brads using diagonal cutters.

Step 4   Dip the brads into superglue prior to working them into the holes. Set them in place with a small brad hammer, as shown next.

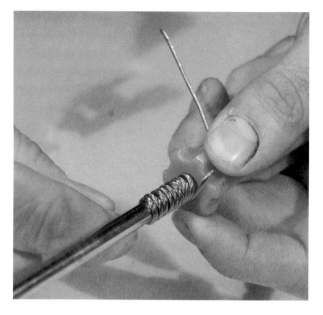

Note

I prefer to use "gap-filling" superglue for this. Be certain to read all manufacturer's instructions and warnings when using this sort of product. I find it easiest to make a small pool about the size of a dime (or a 5p coin for our British readers) in a cap from a water bottle, or a similar disposal vessel, and then dip the brads rather than attempt to apply glue onto the brads directly from the bottle.

Further Note

It should go without saying that I use a pair of needle-nose pliers when dipping the brads, for fear of gluing my fingertips together. And be certain to wipe off any excess glue using a damp cloth.

Step 5   As a final decorative touch, let's add a few "electrical" coils. Start by cutting a roughly 3-foot-long piece of braided copper wire (or whatever type of wire you want). Use a large Phillips-head screwdriver shaft or metal rod as the mandrel around which you will wind the coil. Starting a few inches in from the end of the wire, tightly wind the wire around the makeshift mandrel. Estimating the distance the coil will span on the T-pod, I coil the wire until it is about half of the estimated length. Leave an inch or so of uncoiled wire for attachment to the "terminals."

Tip

As mentioned in Chapter 8, a Phillips-head screwdriver is preferable to a slot-head screwdriver to use as a mandrel because the shaft does not flair at the tip.

Step 6   Attach one end of the coil to one of the terminals on the side of the case. Stretch out the coil to a second terminal and attach it in the same manner.

Hint

Be certain to wrap the wire around the terminal post in the same clockwise direction as the knurled nut will tighten. In this way, tightening the nut back on will help to tighten the wire around the post rather than loosen it.

Do not use Loctite to glue your knurled nuts in place if you ever intend to change out your coils—which I have found necessary to do after prolonged use.

Step 7   Repeat this process until you have attached as many coils as you want for decorative effect. Insofar as possible, try to design the layout of your coils between your terminals so that they could all be part of a circuit. If possible, avoid attaching two coils to a single terminal, because this will appear as a "short circuit" even to those who are unfamiliar with the principles of electrical wiring.

Note

Although none of the coils is intended to serve any practical purpose, they help to perpetuate the illusion that the whole unit has some sort of archaic and mysterious function, and that it is not (as we all know it to be) simply a light in an elaborate box that began its life as the starter coil for a Model T Ford.

Stage 11: Attach Hanging Hardware (Optional)

Some of the people for whom we have constructed these quirky little devices have chosen to set them on a shelf or stand them on a table, counter, or desk. If this is your intention, your work is done.

If, however, you intend to wear the unit as a portable chrono-static insulator, you have a few more steps to complete.

Note

Although this strap is intended to hang the T-pod from a belt, it could easily be changed to a shoulder strap if the reader so desires.

Step 1   Cut a strip of leather (perhaps the leather left over from the front panel), approximately 3/4 inch wide by 5 inches long. Cut that strap into two equal 2 1/2-inch strips.

Step 2   Fold the two straps in half and slide them through two brass "D" rings (available from fabric shops with good notions sections or from a leather craft shop—you could use any type of solid ring here, even key rings). Use a center punch to punch a hole through both ends of each leather

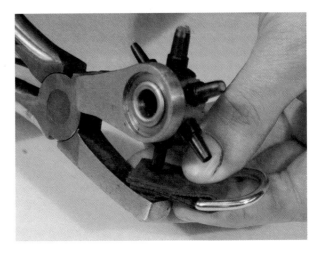

strap, approximately 1/4 inch from the ends. Trim off the corners of the straps.

Step 3   On the sides of your case, mark the position of the screw holes where the strap will be attached. Position them low enough so that you can access the holes on the inside of the box—that is, not into the top panel. (In our case, each hole is about 7/8 inch from the top edge of the case.) Once you have marked your positions, drill out the holes using a 1/8-inch drill bit.

Step 4   Slot two 3/4-inch 6-32 brass bolts through two brass finish washers and then through the leather and the holes in the sides of the case.

Secure these in place with washers and 6-32 brass nuts. Use Loctite or glue the nut to prevent it from working itself loose through the vibration of use.

Step 5   Now we'll add the suspension hardware. Cut two, 3/4-inch × 11-inch leather straps.

Step 6   Slide one end of each strap through a brass snap hook and overlap the opposite end. Punch a hole at that point in each strap, and secure with medium-length double-capped rivets.

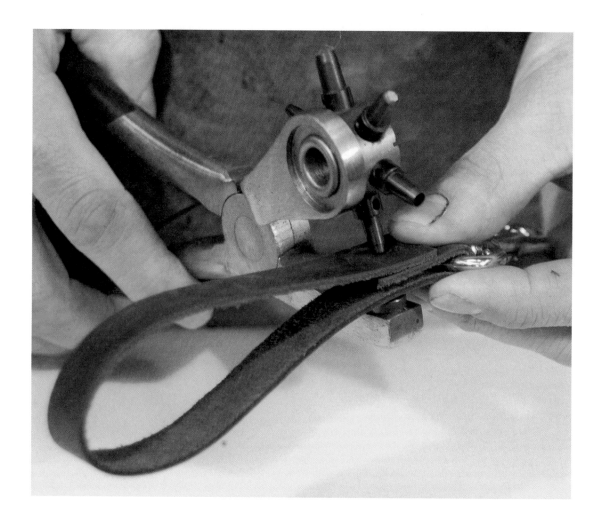

Step 7 Clip your hanging straps on to your mounted "D" rings, and your project is ready to wear.

Final Note

As the whole idea behind creating this wondrous contraption is to have a sort of external pocket in which to contain modern electronic conveniences and necessities such as cellular phones and mp3 players, you may decide to line parts of the interior of your T-pod with foam or felt or velvet or light leather. The lining would mitigate the chances of your electronic device being scratched by any of the hardware inside the case. Only you can determine whether this is necessary based on what size wooden box you chose for your project and what size your electronic device might be. If you do, however, decide to add a lining to your T-pod, be certain to allow access to the switch and battery compartment on your plasma light as well as any other nuts or machine screw heads which you feel may require future adjustment or tightening.

Chapter 10

Altitude Mask with Integrated Respiratory Augmentation as Issued to H.M. Royal Aeronautical Corps

The sun was rising on a clear crisp autumn morning as, 1500 feet above the rooftops of London, Ensign Lloyd sat attentively in the basket of his surveillance balloon, tethered near the banks of the Thames. He was about to pour himself a cup of tea from his flask when the telegraph beeped into life.

"YOU ARE MISTAKEN ENSIGN LLOYD (STOP)" was the response through the telegraph line.

The erstwhile ensign put his eye back tight to the glass. The sleek black coach, with the royal crest clearly visible on its roof, was heading out of Whitechapel and turning down Dock Street toward the Tower of London. But the royal coach was not simply making its way along the fetid thoroughfare: the driver was whipping hard, spurring the horses to a gallop as they raced along.

Lloyd's hand found the transmitter key and tapped out, "WADR SIR (STOP) I AM PRESENTLY OBSERVING RC 04 HEADING FROM WC DOWN DOCK TO TOWER AT HASTE (STOP) WILL RELAY TO SY SPECIAL CONSTABLE AS REQUESTED (STOP)"

Silence followed. Then came the reply: "LLOYD (STOP) STOP (STOP)"

He began to transmit his reply, "AGAIN SIR WADR (STOP) MY ORDERS WERE" and then he realized his telegraph line had gone dead—severed somehow at the mooring. He had only a moment to contemplate how to relay his message when a shudder and jerk shook his surveillance balloon and he jolted skyward. He watched the rooftops of the capital city of the entire empire receding below him as he climbed rapidly skyward, lifted forcefully by the helium expander. Trying not to panic and remembering his training, he strapped on his altitude mask and hooked up the intake valve to the respirator chamber.

Heaven help me, he thought, as the balloon was gripped by the morning wind, dragging him toward the southwest coast and the great expanse of ocean that lay beyond.

–From *Eyes in the Skies: The Possibly True Tale of How a Member of Her Majesty's Geographical Positioning Service Became the Most Wanted Air Pirate and Skywayman in the Empire.* Published anonymously 1889, Tangiers.

*A*n altitude mask with integrated respiratory augmentation? What is it? Is it worn during high altitude adventures? During subaquatic exploration? Or might it be intended for surviving alien atmosphere (at least for the 6 minutes or so before the air supply runs out or the filters clog up) during your next big game safari on Neptune?

Whatever you might claim as the purpose for this device, concoct a story about it. When someone, for example, asks about the gauge, and how you could possibly read it while wearing the mask, simply inform them that next to the hatch of every airship is a mirror, used specifically to check how much air supply you have available when undertaking the daunting task of swinging from one airship to another during a piratical boarding action…or whatever. Whatever fiction you invent, stick to it. Believe in it. Wear your mask with pride and panache, and everyone you meet will agree that it is pretty spiffy.

Project Description

In this chapter, we will be making a breathing assistance mask for wearing at high altitudes. It should be noted, however, that this same design can be adapted for subaquatic, subterranean, or even extraterrestrial adventures. What we will be constructing is (like many of the projects in this book) simply one example of how a breathing mask can be constructed.

With a bit of imagination, some experimentation, and a variety of materials, you can create these masks in many different ways. For the purposes of this chapter, however, I shall assume that the project you are making is absolutely identical in every regard to the one I am making here. That is to say, I will assume that you have all the same materials and tools that litter my workshop and that you are using the same sacrificial mask to make your pattern and the same leather and accessories to build your project.

What You'll Need

To complete the project as described in this chapter, you will need to get your hands on the following materials and tools.

Materials

- Disposable dust mask
- 1 square foot of 5–6 oz. neutral (undyed) leather
- 18, 8-32 × 1/4-inch solid brass machine screws
- 18, 8-32 brass acorn nuts
- 6 small brass double-cap brass rivets and caps, or Chicago screws (optional)
- Scrap of medium-weight suede or even a piece of heavy fabric or craft foam
- 60 inches of 1/2-inch, medium-weight leather strips (for straps)
- 8 medium-size double-cap brass rivets
- 2, 3/4-inch brass "D" rings
- 2, 1/2-inch brass "D" rings
- 2, 1/2-inch square or oval center bar brass buckles
- 2 polished brass 2-inch snap-in tub strainers
- 1/4-inch brass all-tube elbow
- 1/4-inch brass compression cap
- 2, 1/2-inch 6-32 brass machine screw
- 2, 6-32 brass nuts

Tools

- Marker
- Rotary hole punch
- Heavy-duty scissors
- 3/4-inch, oblong slot punch
- Leather dye or similar
- Latex or nitrile gloves

❧ Loctite or white glue

❧ Superglue, epoxy, or barge cement

❧ Utility knife

❧ Ruler

❧ Rivet setter or anvil

❧ Anvil

❧ Bench vise

❧ 5/32-inch drill bit

❧ Power drill (or drill press)

Alternative Tools

❧ Hand punch

❧ Awl

❧ Hammer

❧ Pliers or vise grips

Stage 1: Create Your Pattern

Step 1   Visit your local hardware store and play a bit of dress-up. Ask the helpful staff where you might find the less-than-stylish selection of non-toxic particle and dust masks. Explore the available options for size, fit, and style. The mask you choose will serve as a pattern for the much fancier mask you are about to make.

Note

Prices for a mask can range from a few pennies to $20 or more. As you will likely destroy whatever you purchase in the act of making your pattern, the amount you decide to spend on your sacrificial respirator or dust mask is entirely up to you and your budget. For our purpose in this chapter, I have selected a cheap disposable dust mask. These are readily available and usually come 6 or 12 to a package. (And having a few extras on hand is probably wise if you have never made one of these before.)

Step 2   Cut the elastic or straps off your sacrificial mask. If the mask includes a metal forming bracket across the nose bridge, carefully peel it off without tearing the mask or distorting its shape too much.

Step 3   Use a marker to draw a line directly down the center of the mask, from the highest point on the nose bridge to the base of the chin.

Step 4   Draw a line extending from the part of the mask that corresponds to the end of your nose to a point on the bottom edge, about 1 1/2 inches to the right of the center line. The endpoint should hit around the corner of your mouth and extend down to your jaw line.

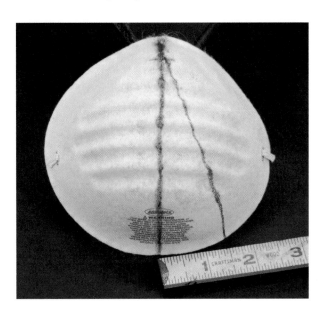

Note

This is not an exact science. No two faces are exactly the same shape or proportion (thankfully), so these measurements are approximations. Alter the pattern to suit your face shape as we go along.

Step 5   Cut your mask in half down the center line you drew in step 3.

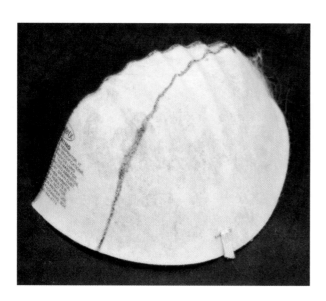

Step 6   Cut along the line you drew in step 4, and hang on to this triangle of material. You should now have three pieces, but you'll be working with the two that look like this:

Step 7   Place both of your cutouts flat onto a piece of paper and trace around them. Leave a little extra space around them on the paper so that you can alter and extend them as necessary.

Note

When tracing out your pattern, ensure that the edge of the triangle, which was once the center line of the mask, remains straight (it is likely to want to curl up or to the side as the result of having been cut from a compound curve).

Step 8   On your paper pattern, mark a small arrow to indicate the top of your larger side piece, because it will be important to know up from down in later stages. Then extend the center line of the triangle piece by 2 inches. This line is labeled "C" on our triangle, as shown in the illustration.

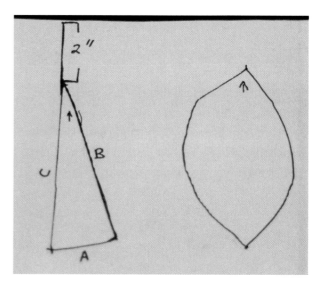

Step 9   Now draw a parallel line 1/2 inch along the outside of line "B," as shown next.

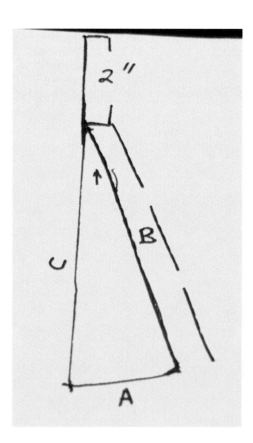

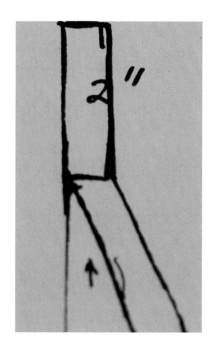

Step 11   Parallel your extension of line "C" 1/2 inch to the right to join up with the new line "B." This part of your pattern should now look like this:

Step 10   Continue the arc of line "A" to meet the new line "B" you drew in step 9.

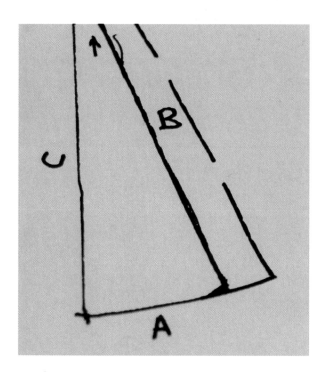

Step 12   Now turn your attention to the mask shape you traced in step 7. First, mark a dotted line about 1/4 inch in from the inner edge of line "D" (which was joined to edge "B" before you cut

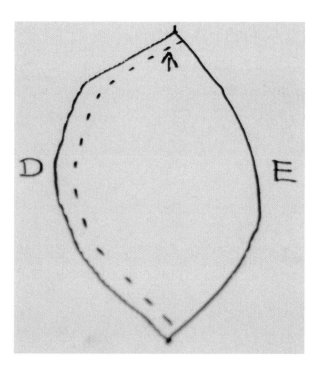

them apart). This dotted line will be the location of your rivet holes.

Step 13 Mark a center point that is 1 1/2 inch outside of your outer curve (line "E")—it should be centered both vertically and horizontally with the pattern shape. We will call this point "Z" as it is the zenith of the curve we are about to create.

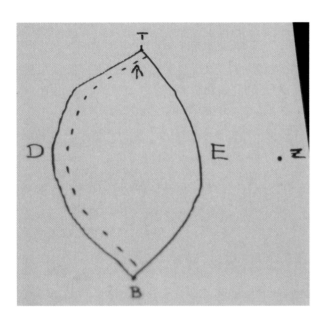

Step 14 Mark a point along curve "E," 1 inch down from point "T" (top) in the illustration. Call this point "X." Now repeat this process to create point "Y," 1 inch up from point "B" (bottom).

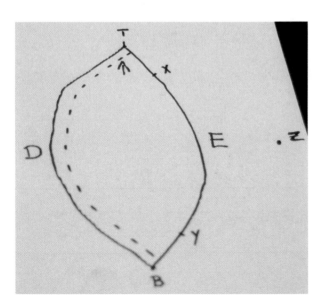

Step 15 Now create a compound curve that connects points "X" and "Y" and passes through point "Z," as shown in the illustration. We will henceforth refer to this new outer curve as "E." You have just drawn the cheek panel for your altitude mask pattern, and the edge you have just rendered will be the one closest to your ear.

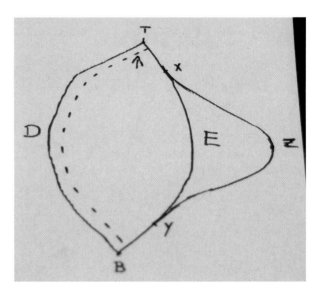

Step 16 Mark out your rivet holes at 1/2- to 3/4-inch intervals, beginning 1/4 inch from points "T" and "B."

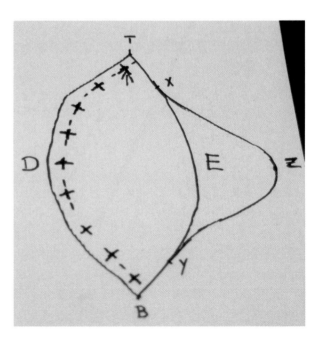

Hint

You want your rivets to be evenly spaced and symmetrical from top to bottom. This might mean that they are not precisely 1/2 or 3/4 inch apart and/or that the spacing may vary slightly between the rivet holes. Use your best judgment and your eye as a designer to make this determination.

Step 17   Cut out both pattern pieces.

Step 18   Punch out your rivet hole locations using a rotary hole punch set on its smallest setting.

Step 19   On the triangle-shaped pattern piece, mark a faint line all the way up and about halfway between the outer line "B" and the inner line "B."

Step 20   Match up edge "D" with inner line "B," starting at the bottom point. The rivet hole should fall exactly 1/4 inch up from the end of the faint line you have just drawn. Mark the location of this rivet hole on the triangle-shaped pattern.

Rotate the larger cheek panel slightly so that edge "D" remains in contact with inner line "B," and mark the position of the next rivet hole up along the faint line.

Continue doing this from one end to the other until all the rivet holes have been traced on the triangular piece.

Note

If your rivet holes are at exact measured intervals (such as 1/2, 5/8, 3/4 inch, or whatever) you can simply mark those out on this faint line with a ruler. But if the rivet holes are spaced at varying distances based on symmetry, as in our case, I would recommend the process as explained in step 20.

Step 21   Punch out the rivet holes in the triangle-shaped pattern.

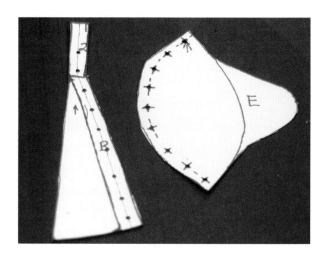

Step 22   Get a fresh piece of paper. Mark a straight line, 6 to 8 inches long, down the center of

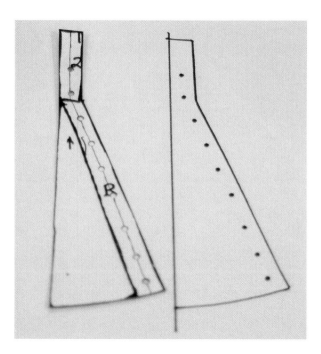

the page. Align edge "C" with the line you have just drawn and trace the profile of your triangular front panel (including the locations of the rivet holes).

Step 23   Flip over your pattern on the paper, and again match up edge "C" to the center line and trace out a mirror image of the profile of the pattern with rivet hole locations. You should now have something that vaguely resembles a laboratory beaker and looks like this:

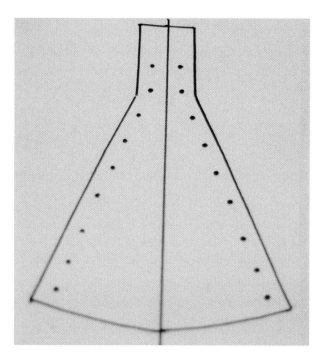

Step 24   Cut out your new beaker pattern (as we will now call it) and punch out the rivet hole locations. Do not cut along the center line.

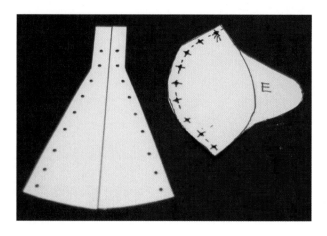

Note

You would do well to save these patterns. You will likely use them again. Of course, you have only one face and might, therefore, need only one mask. But sooner or later, you will be inspired by a new design, you'll want to coordinate it with and accessorize a new outfit, or some friends will plead and cajole until you make one just like it for them.

Note

If you do not have a leather supplier nearby, you can use leather salvaged from old bags and purses of a suitable weight. Alternatively, I have seen this style of mask constructed from stiff upholstery fabric, sheets of craft foam, and even neoprene (talk about a sweaty face)—all to varying degrees of effectiveness. If possible, however, use leather.

Stage 2: Cut Out Your Mask

Step 1   Choose your base material.

Our mask will be made out of 5–6 oz. neutral (undyed) leather. Through years of experience (which is to say, trial and error—mostly error), I have found that this leather weight works best for altitude mask projects. You shouldn't need too much of it—about 1 square foot of leather will do. At the leather shop, check out the scrap bin for pieces, rather than buying an entire hide (unless, of course, you are intending to make dozens of masks).

Step 2   Use a marker to transfer your pattern onto the leather. Because any piece of leather has a finish side and a flesh side, trace out your patterns (including the rivet hole locations) on the flesh side of the leather. This way, extraneous marks will not be visible on the outside of the finished piece.

Step 3   Flip over the cheek pattern and trace its outline (and corresponding rivet hole locations) again. Be sure to label the cheek panels "Right" and "Left" and mark the top. I have used a small arrow pointing up for this purpose.

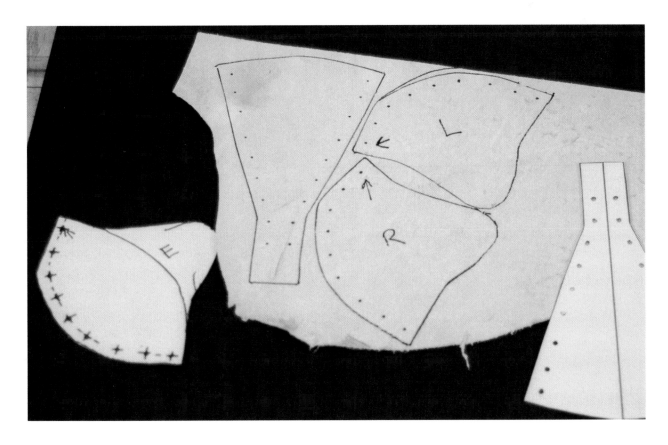

Note

The cheek patterns are unlikely to be perfectly symmetrical, which is why it is important that you keep track of which is the right and which is the left cheek. Attaching them incorrectly, on the wrong side, or upside down will not only mean that you will have a tough time getting your rivet holes to align, but you will also have a difficult time getting the mask to fit properly or conform to the curves of your face.

Step 4   Cut out the leather panels using a pair of heavy-duty scissors.

Note

Some leather workers might suggest using a craft knife for this purpose, but the combination of internal and external curves makes this a difficult option for the novice. If you believe that your kung fu is strong enough to handle a blade for this task—it is your project, after all—and you are happy with choppy edges (and your insurance is up to date), then so be it.

Step 5   Punch out your rivet holes using a rotary hole punch set on its second-largest hole size.

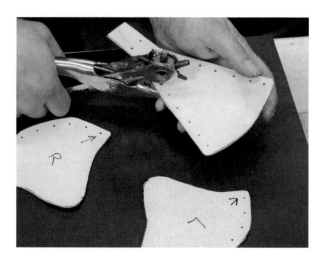

Hint

In the interest of both comfort and appearance, I recommend that you slightly round off the corners of your mask. This should also help mitigate the chance of the corners curling.

Stage 3: Add Breathing Slots

Step 1   We need to add some ornamental breathing slots on the front (beaker) panel of the mask. I like how these make the front of the mask resemble an old Studebaker grill. To begin, get out the beaker pattern and mark out where the slots should appear.

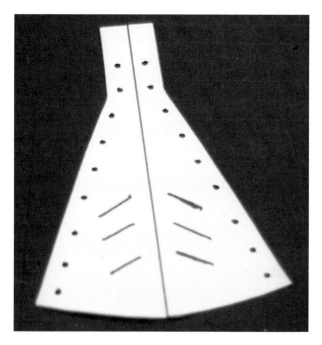

Step 2   Using a 3/4-inch, oblong slot punch (typically used in making buckle slots), punch holes in the pattern at the breathing slots. I have several of these punches in different sizes that I picked up from Tandy Leather. If you can't find these locally, you can get them online.

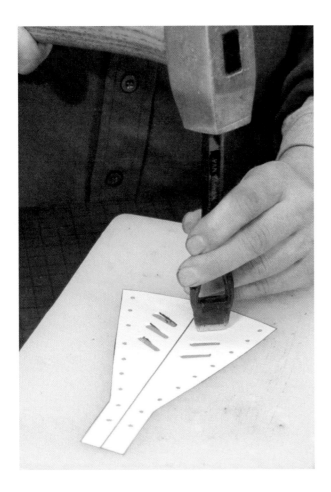

Note

When punching out slots using this method, be certain to use a backing board (lest you damage your work surface). A scrap of plywood, a plastic or wooden kitchen cutting board, or even an old phone book will suffice for this purpose.

Stage 4: Dye the Leather

Your choice of color is, of course, entirely dependent on your design and personal esthetic. In this case, we'll use two different dyes: a medium brown dye and a buckskin dye. Leather dyes should be easily available from most hobby shops or, of course, leather suppliers. You can also check online.

Hint

Wear latex or nitrile gloves when working with leather dyes, lest you risk becoming dyed yourself for several days. Remember that you, too, are covered in undyed leather (tattoos and self-tanning notwithstanding).

Hint

If you do not have a slot punch and do not fancy purchasing one, consider punching a large hole at either end of each slot and using a sharp craft knife and a straight edge to make parallel cuts from one circle to the other. If you are attempting this method, work carefully and proceed slowly. These slots, after all, will be front and center on your finished mask, and any inaccuracy or variation in your slot lines will be very apparent.

Step 1   First, of course, read all of the instructions that came with your leather dye or stain. Then use the medium brown dye to color the beaker-shaped leather and the buckskin dye to color the two cheek panels.

Note

Any dye that will soak into the leather should work for this purpose. You can experiment with wood stain, ink, or even markers. I recommend that you do not use standard paints, however, which would sit on the surface of the leather and be prone to cracking and flaking as the leather is worn, used, and flexed.

Step 3   Use the pattern to transfer the slot locations onto the finish side of the front leather beaker panel, and then punch out the breathing slots in the leather.

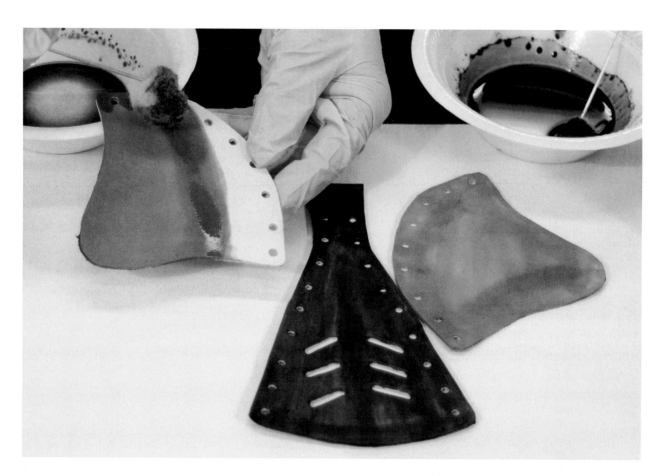

Step 2 Wait the recommended amount of time for the dye to dry thoroughly before you begin Stage 5.

Stage 5: Assemble the Mask

Step 1 Arrange your panels such that the rivet holes on the cheeks and beaker panel line up. Be certain that the front beaker panel overlaps and is on the top (outside) of the cheek panels.

Step 2 Get yourself 18, 8-32 × 1/4-inch solid brass machine screws and corresponding 8-32 acorn nuts. (These should be available at any well-stocked hardware store.) Slot the machine screws through the rivet holes from the inside out (flesh side to finish side) and secure them in place using the acorn nuts.

Note

Instead of screws and acorn nuts, you can use small brass double-cap rivets (as we have used in other projects in this book). This screw and acorn method, however, requires fewer specialized tools (such as a rivet setter or anvil), and also means that the whole mask can be assembled and disassembled to change parts or panels as desired. Another alternative to these screws and acorns would, of course, be to use Chicago screws (available through most leather suppliers) if you are lucky enough to find really short ones.

As we are in the process of converting three flat panels into a single mask comprising compound curves, do not feel shy about stretching or bending the panels so that the rivet holes align as you proceed.

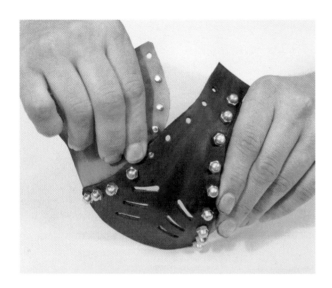

Hint

If, of course, you are happy with the design of your mask, you can use a dab of Loctite or white glue to secure the acorns in place. Be aware, however, that this will limit your ability to disassemble your mask to interchange parts or panels.

Stage 6: Create a Nasal Liner

Assuming you have followed along precisely thus far, your project should closely resemble the following images (front and back):

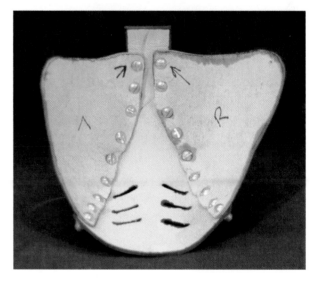

Step 1 Fold the extra length at the top of the beaker panel over to the inside and fix it in place using some superglue, epoxy, or barge cement. This will not only give the top edge of your mask a more finished and professional look, but it should also make the mask far more comfortable to wear.

Step 2 We will now mark out and cut out a nasal liner using a scrap of medium-weight suede or even a piece of heavy fabric or craft foam. Cut out a rectangle of roughly 2 1/2 × 1 1/2 inches.

Step 3   The liner should just cover the screws at the top of the mask, where the mask rests on the bridge of your nose. Glue the nasal liner in place with superglue, epoxy, or barge cement.

Caution

Do not rivet or stitch the nasal liner in place, because the rivets or stitched seams will be even more irritating to your nose than the stuff which you are attempting to cover.

Stage 7: Add the Straps

Your mask needs straps to hold it onto your face. Here, I will show you how to put a buckle on a strap and make the tongue (the part with all the holes for adjusting the fit). You might need these skills for other projects, and these are good things to know anyway.

Note

Make sure you are measuring away from the ends of the pieces for all these holes.

Step 1   Start with two pieces of medium weight leather, cut into 1/2-inch strips, each about 30 inches long.

Note

For this purpose, old belts are frequently good sources of leather.

Step 2   From each of those strips, cut one 12-inch-long piece (a total of two pieces). Write a "B" on the flesh sides of both pieces.

Step 3   Cut two 15-inch-long pieces from the strips, and label these "T" on the flesh side.

Hint

Always remember to mark your leather on the back side whenever possible. Remember what I said about trimming off the corners too.

Step 4   Finally, cut two 3/4-inch-wide pieces, each 3 inches long. When you finish cutting all the pieces, you should have six pieces total:

Step 5   Use a rotary punch to make a rivet hole 1/4 inch from each end of both of your 3-inch pieces. These are your ring straps.

Step 6   Mark a hole near the "E" edge of your mask on the cheek panels, about 3/4 inch from point "Z" (near the ear). Using your rotary punch tool, punch out a hole for a rivet.

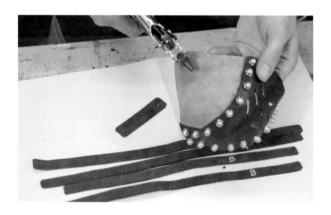

Step 7   Slide the post of a medium-size double-cap rivet through the finish side of one of the end holes on the ring strap, and then through the hole you just punched in the mask cheek panel, making sure that you go from inside to outside.

Step 8   Slide a 3/4-inch "D" ring onto this ring strap, and then fold the strap over and secure the

other end over the rivet post with a cap and tap. Do the same to the other ring strap and cheek panel.

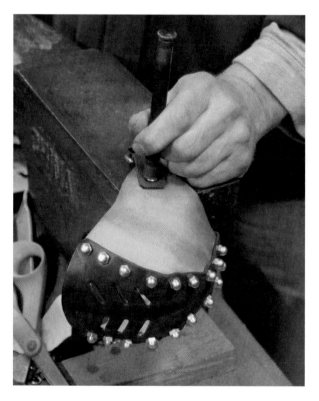

The next few steps walk you through adding a 1/2-inch buckle on the end of a 1/2-inch-wide leather strap—as discussed in Chapter 5. This bears repeating, because you will tend to use these a lot.

Step 9   Grab a strap labeled "B." Place a mark for a hole 1/8 inch from one end—this will be the buckle end. Measure 3/4 inch from that mark and make another mark. Measure 1/2 inch from that mark and make another mark; then measure 1/2 inch from that mark and make yet another mark. Then, you guessed it, measure 1/2 inch from that and make a mark. Now make one more mark 3/4 inch from the very last mark. You should have made six marks on this end of the strap.

Repeat for the other B strap.

Step 10   Set your rotary punch to the second smallest sized hole and punch holes at the six marks you just made.

Step 11   See those two holes in the middle of the six you just made? You are going to connect them with two parallel lines extending tangentially off the two holes. To cut these slots, you can use the oblong slot cutter you used back in Stage 3, or you can use a craft knife with a sharp blade and a ruler or straight edge.

If you use a knife, here's how you make the cuts: Place a strap horizontally and flat on your worktable. Place the straight edge against the bottom of the two middle holes, so you can see the holes. Working very slowly and carefully, cut a straight line connecting the two holes. Rotate the strap 180 degrees and place the straight edge along the bottom of the holes and cut across like before. You have now turned the two

holes into a slot. This end of the strap should look like this when you are finished:

Step 12   Take a good look at your buckles. The type of buckle I recommend for this project has three parts: the prong that sticks through the holes on a belt, the bar on which the prong swivels, and the frame that goes around the outside. On many buckles, one section of the frame is a bit thicker than the other. I will call this the front of the frame. So now we have all that out of the way, we can get on with putting the buckle on the end of the B strap.

Step 13   Holding the buckle horizontal, put the buckle end of the B strap through the back end of the frame of the buckle. Stick the prong of the buckle straight up and into the slot you cut into the strap. Put the end of the strap down through the front part of the frame so it wraps around the bar.

Step 14   Now there is, of course, more riveting. Place the post of a medium double-cap rivet through the good side of the hole just to the front end of the slot. Wrap the strap around the bar and put the tip of the stud through the back side of the hole on the other side of the slot. Put the riveted leather over the flat part on your anvil or the top of a vise. Add a small cap and whack it.

Step 15   From the other end, slide a 1/2-inch "D" ring up the length of the strap to the rivet just behind the buckle. Make sure the round part is up and the flat part is sandwiched between both pieces of the leather strap. Pass another medium-size rivet stud up through the hole in both pieces of leather behind the "D" ring, trapping it between the two rivets. Time to cap and whack. Again use a small cap. Now you have a buckle on the end of a strap. It should look like the following image.

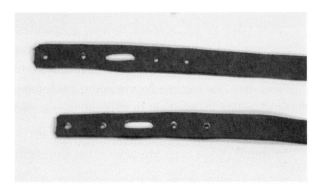

Step 16 Use your sharp marker to mark the location of two more holes in the buckle strap. Mark one hole 1/8 inch from the other end. Measure 1 1/2 inches from the center of the end hole and make another mark. Punch two rivet holes at these marks. These two holes are for the "D" ring end of this strap.

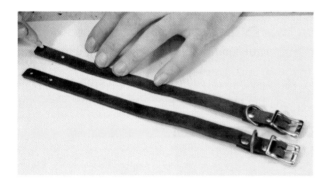

Step 17 Thread the "D" ring end of the B strap through the "D" ring on the left cheek panel from the outside, wrapping around the "D" ring. Line up the two end holes, add a medium rivet stud through the finish side of the end hole, up through the back side of the other hole on this end, and now, you guessed it, cap it and whack it. Repeat for the second buckle strap—*also* on the left cheek "D" ring.

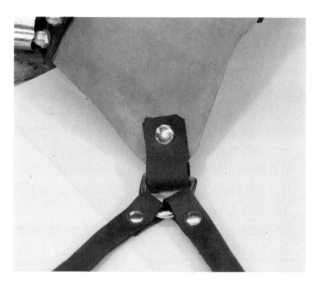

Step 18 On to the T strap (see the image below). Make a hole 1/8 inch from the end—does not matter which end. Now make another hole

1 1/2 inches from the center of the last hole. This is now the "D" ring end of this strap. Measure 3 inches from the hole you just made and make a mark. Measure 3/4 inch from there and make another mark. Keep doing this—measure 3/4 inch and make another mark—all the way to the end of the strap. Set your punch to the second to largest hole size, and punch a hole at each of the marks you just made.

These holes are larger than the rivet holes to make it easy for the prong of the buckle to go into them. Cut the end just after the last hole to a point to make it easier to insert into the buckle.

Step 19   Attach the T straps onto your right cheek panel "D" ring (in the same way you attached the buckles onto the left cheek panel "D" ring), and your straps are done.

Go ahead and try on the mask for size. Keep in mind that the top strap is designed to go over the ears and around the back of the crown of the head, while the lower strap is designed to reach around the neck at the base of the skull.

You could stop here and have a fully serviceable mask, and I even promise not to laugh at you when I see you wearing it. Nary a chuckle or a smirk. Cross my heart. You know you can't stop, though, don't you? Yeah, I couldn't either.

Stage 8: Add Ornament

What we have at this stage is a perfectly suitable leather altitude mask. But, at the moment, it is fairly plain and more like a surgical mask than something to be used at high altitudes as a respirator or as a subaquatic scuba system or alien atmospheric adapter. So let us focus on a bit of fanciful steampunkery.

To find ornaments for this mask, I visited my local hardware store and scoured the plumbing and housewares selection. Within minutes, I found and purchased a few pieces to use for respirator vents on the sides of the mask. That said, I have come to know the available selection and location of most items in the hardware center almost intimately by this stage.

The pieces I have selected are as follows: two polished brass 2-inch snap-in tub strainers, a 1/4-inch brass all-tube elbow, and a 1/4-inch brass compression cap.

Step 1 I will assume, again, that you have found and selected ornamentation components identical to mine. Place the mask over a support of some type, such as the corner of your worktable, cheek panel facing up. Position one of the snap-in tub strainers at the center of one cheek panel. Press the strainer into the leather, and the flanges that extend off the bottom side of the strainer will leave dented impressions in the cheek panel.

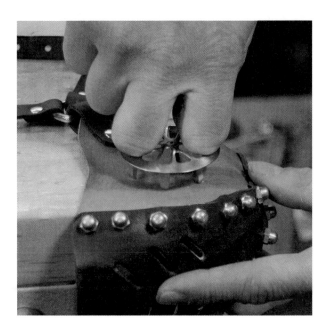

Note

Of course, the degree to which the dents are visible is entirely dependent on what sort of leather you have used. Feel free to clarify the impressions with a pen or marker, because they will be covered by the strainer rim anyway.

Step 2 Using your rotary punch set on the largest setting, punch out holes at the marks left by the flanges.

Note

If you have trouble reaching the innermost holes with your punch, you might need to use a hand punch and hammer, or perhaps an awl. Bear in mind that since these holes are intended to accommodate the flanges of the strainer, you could—at least in this instance—use a sharp blade to pierce small slits through the cheek panel in place of punching holes.

Step 3 Insert the flanges through the holes you have just punched and use a marker to draw in the slots in the strainers that line up with the position of the flanges. You will be cutting out these slots.

Step 4 Remove the strainer. Using a craft knife or the slot punch you used when making the buckle slots, cut six slots in a radial pattern, with the holes you just punched serving as the outer points of each slot. The cheek panel should look like the image at the top of the next page when you're done.

Step 5   In order to increase the illusion of depth, take a good black marker or some black leather dye and color in the triangles which exist between your slots. And don't forget to color in the interior edges of your slots.

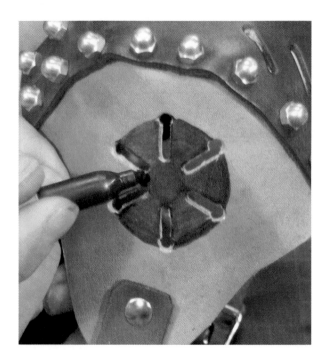

Step 6   Punch a hole in the center of the circle to align with the center hole of the tub strainer.

Step 7   We will now add further ornament to the strainers (lest it simply appear that we have a drain on each cheek). On one side, we will add the all-tube elbow and cap. But first, we need to enlarge the center hole of the strainer. Carefully drill it out using a 5/32-inch drill bit in a power drill.

Step 8   Clamp the 1/4-inch brass compression cap tightly in a vise or hold it between a large pair of pliers (or even vise grips, as in the photo on the top of the next page). Then drill a corresponding hole of the same diameter (5/32 inch) in the cap.

Note
Because of the way in which these caps are machined, the center point should be clearly identifiable.

Step 9   Put your strainer back in place on one cheek panel. Bend the flanges flat on the interior of the mask, radiating outward from the center. If your fingers are not strong enough to bend them flat, use pliers or a small hammer.

Step 10   Drop a 1/2-inch, 6-32 brass machine screw through the hole of the compression cap from the inside out. Use a drop of superglue to hold it in place. Wait for the glue to dry.

Step 11   While you are waiting for the glue to dry, work on the other cheek panel and the second strainer.

Step 12   Return to the cap from step 9 and slot the machine screw through the center hole of a strainer. Pick a side, any side. Secure the cap and screw in place using a 6-32 brass nut. Be sure to use a small amount of Loctite or white glue to help prevent the nut from working itself loose.

Step 13   Thread the all-tube elbow into place securely within the cap.

You could, of course, finish here. And it is often said that less is more. But if this is true, then think how much more, more would be. Let us take it at least one step further.

Remember those gauges we made in Chapter 6? Remember how I mentioned that they are incredibly useful for augmenting and ornamenting other sorts of projects? Well, let's use one now. If you have not made any of these, perhaps you should consider going back and making some now. Alternatively you may have found other parts or components that you want to use to adorn your mask. Your designs are, of course, entirely your own—as well they should be. But for our purposes here, I am going to use one of those gauges. To paraphrase virtually every cooking and craft show on television, here's one I made earlier (see next column).

Step 14   After selecting a flush-mounted gauge, simply slot the integrated machine screw through the central hole of the other strainer and secure in place with a 6-32 brass nut and a dab of Loctite or white glue.

Step 15 I probably don't even need to mention this, but for the purposes of completeness, if your screws, bolts, or nuts extend too far into the mask and are scratching your face, you should cut them off and file them down using metal files or a rotary tool.

Note

You could of course also choose to line the interior of your mask with thin suede or fabric. I do not think it is necessary, but this is entirely a matter of personal preference. If you do decide to create a lining, you can use the patterns you created earlier and secure the lining in place with fabric glue or leather glue (let the glue dry completely before trying to breath through the mask), by stitching the lining to the outer edge, using screws and acorn nuts to hold it fast, or by any combination of the above. It is, after all, your mask...so you get to design it.

There you go. It is as simple as that. Now that you know how to make this style of mask, play around with the pattern and design. Try making one out of other materials. Try different sorts of ornamentation. Perhaps you could even try making several of these in different color combinations to coordinate with your Steampunk wardrobe (or your wallpaper—who knows). Play around with it and have fun. And if you see me out and about at a gathering somewhen, do please come up to me and show me what you have made.

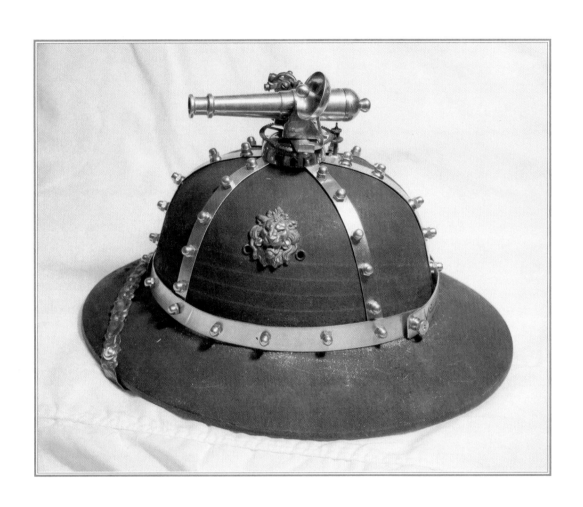

Chapter 11

Voortman's Armoured Pith Helmet, from London's Finest Purveyor of Defensive and Deflective Haberdashery

Ah yes, as I was saying, unlike the *Tyrannosaurus rex*, the *Victoriasaurus regina* has rather long forelimbs, each with a single wicked claw, that it keeps coiled close to its body, poised to thrust out as though they were spring loaded. Nasty piece of business that. We of course had no idea of this till we ran across one on our hunt for the rare Grey Footed Marmoset—the key factor in the production of the vaccine for some obscure ailment I have forgotten about. A shame the nasty blighter's first swipe took the top of old Lord Floggenhall's head clean off.

While the others of my party were busy fleeing for cover and/or soiling their unmentionables, I found myself inexplicably, and somewhat foolishly, reaching for my 97.62 Caliber Bronto-Blaster Express. The first four shots managed to get the beast's undivided attention. Unfortunately, it was time to reload. Good thing I was wearing my Voortman's Armoured Pith Helmet. The thing's second swipe was brilliantly deflected by my helmet, leaving my cranium mercifully intact. At this point, the helmet's Crest Mounted Proximity Auto-Cannon had locked onto the beast and provided sufficient cover fire for me to finish reloading. The bloody thing managed to devour three of our native guides and the rest of Old Lord Floggenhall before I could bring it down. Who knew it had three hearts and a brain the size of a walnut? I plan on wearing my Voortman's Armoured Pith Helmet on every expedition.

—From *A Testimonial by Lord Archibald "Feathers" Featherstone*, published in Voortman's Catalogue of Defensive and Deflective Haberdashery

In my humble opinion, no item of clothing says "I am on an adventure!" quite like a pith helmet. Pith helmets take a close second to brass goggles as a tribal identifier for the entire Steampunk subculture. That being said, there is no reason that you need to be content with a commercially available pith helmet, any more than you should be content with a cheap pair of plastic goggles.

Project Description

In this chapter we will take an inexpensive, safari-style pith helmet and modify it to look even more interesting and adventurous. Individuality in design is of fundamental importance, so this chapter will show you how to customize your pith helmet to make it your own. Furthermore, the principles and techniques described here can be used or adapted to modify and customize a wide variety of wearable accessories.

What You'll Need

To complete the project as described in this chapter, you will need to get your hands on the following materials and tools.

Materials

- Pith helmet
- 6, 12-inch × 1/2-inch brass strips, 0.032-inch thick
- 36, 8-32 × 1/4-inch brass machine screws
- 48, 8-32 brass acorn nuts
- 12, 8-32 × 3/8-inch brass machine screws
- 3, 12-inch × 3/4-inch brass strips, 0.032-inch thick
- Toy cannon, preferably brass
- Small alarm clock bell
- Large alarm clock bell
- 3, 8-32 × 1-inch brass machine screw
- 3/4-inch-long piece of 1/8-inch aluminum tubing
- Threaded brass sphere
- 3, #8 brass nut
- 4, 6-32, 3/8-inch brass bolts
- 4, 6-32 brass nuts
- Decorative hardware (brass lion mask)
- 24–36 inch textured brass strap

Tools

- Safety glasses
- Spray primer
- Spray paint
- Sharp black marker
- Sharp white or silver marker
- Aviation snips (or heavy scissors)
- Sandpaper or metal file
- Spring-loaded center punch
- Drill press
- Power drill
- 5/64-inch drill bit
- Clamps

Alternative Tools

- Fabric or leather dye
- Band saw
- 1/8-inch awl

Stage 1: Paint the Helmet

Step 1   Select and purchase a pith helmet. Older pith helmets can be found at flea markets, antique shops, and occasionally even army-navy surplus stores. A few companies make and sell good-quality reproductions of Victorian and Edwardian pith helmets, though most of them substitute dense cardboard in place of the traditional pith. Any of these reproductions (readily available online) will suit our purposes well and may be preferable to those of you who dread the thought of mutilating an antique.

Note

There are a number of manufacturers out there who are producing cheap soft foam or plastic pith helmets for costume use. While these cheap alternatives may serve well for a photograph or a child's birthday party, for the purposes of modification they should be avoided at all costs. We will require the rigid structure of the harder helmets to realize this project.

After you have selected and purchased a pith helmet, you need to spray primer on the surface of your helmet to get it ready to paint.

Note

Oh, and, do make sure the helmet fits (or can be easily adjusted to fit). No amount of armor added to the pith helmet will protect it from your wrath and rage if you discover it doesn't fit after you have dedicated much careful and diligent work.

I have chosen to use a dark primer, because I intend to make this helmet black (which will contrast nicely with the brass reinforcing strips I will attach to it), and I want it to appear somewhat ominous. You could, of course, use a pale primer and a tan or khaki finish, but these hues can be so tonally similar to the brass finish of the hardware and ornament that the latter is rendered virtually invisible. Read the manufacturer's instructions on your primer packaging, and then paint the helmet.

Step 2 After your primer has dried thoroughly, you can spray paint your pith helmet with a finish color.

Note

If your pith helmet has a fabric covering, you can use fabric dye or even leather dye in place of paint; paint should, however, work equally well.

Stage 2: Add Vertical Reinforcement Straps

Most, if not all, pith helmets have six seams running vertically from the crown to the inner edge of the brim. In this case, I will disguise these seams with brass reinforcing straps and rivets—which should look pretty cool.

Step 1 You can find strips of 1/2-inch-wide brass strapping in hobby shops and craft centers. It should be available in various thicknesses and gauges. The straps I am using are 0.032-inch thick. The straps normally come in 12-inch lengths.

Step 2 Place one end of one strap tight against the rim of the helmet and flex the strap along the curve of the dome to the button on the crown. Use a marker to mark the strap where it meets the button. Do this for each of the helmet's seams—six in all.

Hint

You might find it useful to mark each of these straps on the back side. Number them 1 to 6 starting from your front seam and write corresponding marks carefully onto each of the helmet seams using a silver or white marker.

Step 3   Cut each strap to the desired length using aviator snips or a band saw (if you are working with a thinner gauge brass, you can probably cut it with good heavy scissors).

Step 4   Clip, sand, or file off the corners of the straps.

Step 5   Use a marker to mark the position for attachment holes, 3/4-inch down from what will be the top of each strap and 1 inch from the bottom. Decide how many attachment points you want to position between the two you have just marked, and mark the locations for those as well. I have included three more attachment points in addition to the two initial anchor points (top and bottom) for a total of five holes.

Hint
When designing the number of attachment points, odd numbers tend to work better than even numbers.

Step 6   Use a spring-loaded center punch to dimple the hole points. This is an important step: it mitigates the possibility of the bit "walking" across your brass stock as you drill. Don your safety goggles, and then drill or punch out the holes in your brass straps. A power drill or drill press fitted with a 5/64-inch drill bit works well.

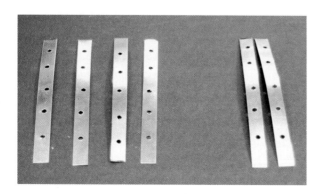

Caution
Be sure to drill these straps out individually and ensure that the strap is securely clamped down before you begin drilling. Failure to secure the strap prior to drilling runs the risk of the drill bit binding in the brass and turning your brass reinforcement strap into a spinning samurai sword of finger-shredding doom.

Step 7   Hold each strap against its corresponding seam, and transfer the hole locations using a sharp marker. Because my helmet is painted black, I use a silver marker to transfer these holes. Do this for each strap in turn.

Step 8   Drill or punch out the holes in your pith helmet. Use a power drill fitted with the same 5/64-inch drill bit.

Note

I have chosen to use some finish washers on straps 1 and 4 to add a bit of detail. To accommodate these washers, therefore, I have used 3⁄8-inch machine screws in place of the 1⁄4-inch machine screws on these two straps.

Stage 3: Add Horizontal Reinforcement Straps

At this stage, we'll add horizontal reinforcement straps onto the helmet. These will not wrap around the entire brim of the helmet, because we'll be adding a maker's plate in steps 10–12 to include as part of the strap.

Note

The lengths I use here fit my helmet. The lengths of straps you need to use will depend on the size of the helmet and the size of the maker's plate. So read through these steps before you begin and make adjustments as necessary.

Note

If your pith helmet is made of pith or cork, you can probably punch out these holes using an awl. This is especially desirable if your pith helmet has a fabric lining and/or outer covering that would likely bind or twist and wrap itself around the spinning drill bit. My helmet is constructed of some form of pressed, dense cardboard or fiberboard (many helmets, however, are made of fiberglass or even various types of plastics).

Step 9 Although you could now rivet each of your reinforcement straps to its corresponding seam using either double-cap rivets or pop rivets, I have a different effect in mind. As with most of these projects, I have to assume that the parts you are working with are identical to mine. I suggest, therefore, that we use 1/4-inch, 8-32 brass machine screws and secure them in place with 8-32 brass acorn nuts. Use a dab of Loctite or white glue when securing the acorn nuts in place, lest they work themselves loose though use, vibration, or wear.

Step 1 Take two 12-inch lengths of 3/4-inch-wide brass straps (available from decent hobby shops and craft centers). Mark a spot for a hole, 1/2 inch from one end of each strap. Use a spring-loaded center punch to dimple it before drilling it out using a 5/64-inch drill bit. File or sand off the corners of these brass bands slightly to mitigate snag points.

already in place on the helmet). Dimple the location with a spring-loaded center punch and then drill it out using a 5/64-inch drill bit.

Step 2   Mark a position for a hole 3/8 inch from the bottom edge of vertical strap 1 (which is

Step 3   Now use a longer (3/4 inch to 1 inch) machine screw with a nut to position the two horizontal strap bands on the front of the pith helmet, as shown at the bottom of this page.

Step 4   Bend the band straps around the brim of the pith helmet, marking the location of each of the vertical straps. Then mark hole spaces on the horizontal straps at even intervals between the vertical straps.

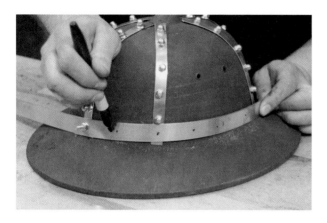

Step 5   Keeping track of which band is which (mark them R and L on the inside, and mark the same on the helmet), remove the two bands from the helmet. Punch and drill out the rivet holes using a 5/64-inch bit.

Step 6   Screw the two horizontal straps back in place at the base of strap 1, and mark the position of the holes from the bands onto the helmet's straps and rim.

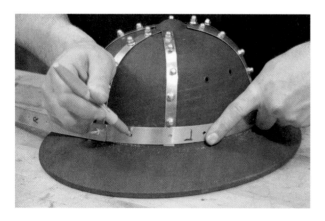

Step 7   Remove the two bands again. Punch and then drill out the holes you have just marked around the brim.

Step 8   Now replace the front (strap 1) bolt with an 8-32 × 3/8-inch brass machine screw, and then lock both ends of the horizontal bands onto it with a brass finish washer and an 8-32 brass acorn nut.

Step 9   Proceed around the perimeter of the helmet, attaching the band using 8-32 × 1/4-inch brass machine screws and 8-32 brass acorn nuts. Use two 8-32 knurled nuts to hold the ends in place.

Step 10 Now you'll measure and cut a third band from the 3/4-inch brass band stock. This band should be long enough to span the gap and barely cover the ends of the two horizontal bands you just attached to the helmet.

Step 11   Punch and drill mounting holes 1/2-inch from either end of this third band, and then trim, file, or sand off the corners.

Step 12   This third band is going to be a maker's plate that overlaps the ends of the two horizontal bands. I have engraved a maker's tag onto the band. (In this case, I have actually written it on with a fine-point marker prior to engraving, as I fear the engraved text will not show up well in the photographs for this book.)

Bend the plate just a bit to roughly match the curve of the helmet. Remove the two 8-32 knurled nuts from the ends of the straps. Place the holes in the plate over the bolts, replace the knurled nuts, and tighten them down with your fingers.

Note

If you are intending to create a maker's label, you will have to decide whether it will be your own label as a maker (which in my case would be Brute Force Studios) or whether you will be creating a fictional manufacturer. In this project, I am using the fictional manufacturer that I listed in the "testimonial" that opened this chapter. I engrave, therefore, "Voortman's Armoured Pith Helmet, London – Bombay – Neptune".

Stage 4: Mount a Cannon and Other Ornaments

In the fictional testimonial that opened this chapter, I described that the pith helmet was not merely armored and protective but that it also possessed a "Crest Mounted Proximity Auto-Cannon." If you are happy to simply armor your helmet, then you can stop here, but let's take things a bit further, shall we? Let's give this helmet some bite.

I'm going to mount a tiny cannon on the crest of my helmet, along with a proximity sensor, a satellite dish, and a few gears. If you examine the contents of this photo, you will see the assortment

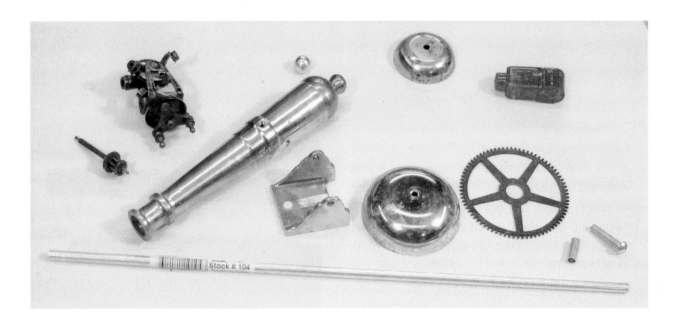

of hardware I have chosen to complete my version of this project.

The degree of ornament you choose to add is entirely up to you. This design is, as with every project in this book, simply one example of what can be done with a bit of imagination, some experimentation, and some patience.

That said, your design and selection of hardware, of course, might be different from mine. Then again, you might be perfectly content with the pith helmet described and shown here. If you want your pith helmet to be purely defensive and protective, you have only to cap off the top with an old alarm clock bell or similar (see Figure 11-1). If, on the other hand, you want for your helmet to be counter-offensive as well as defensive, then by all means do carry on.

Some possible ways to finish this off are shown in Figure 11-1.

Step 1   Undoubtedly in your travels and searches, you have accumulated a fair number of small toy cannons. If not, get hunting. Examine your selection of assorted artillery and choose one that seems to be of a suitable size and scale for the helmet you are designing.

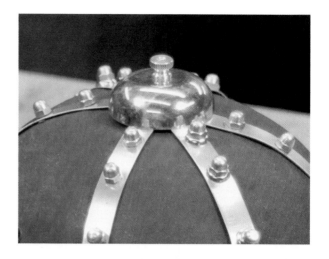

Figure 11-1   *Bell top finish, spike top finish, governor finish*

Step 2   Deconstruct the cannon until only the barrel remains (or the barrel and trunnion mount if your design calls for it). I have decided to attach the trunnion mount from the toy cannon on top of the helmet, so when I cut it apart, I was certain to make it as flat as possible.

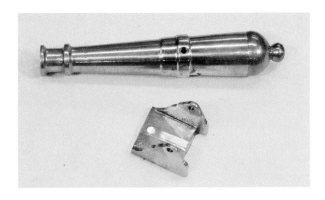

Step 3   Select a large alarm clock bell from your assortment of remnant clock parts. We will use this bell as a base for the gun mount and to cover the existing button on the crown of the pith helmet and the top ends of the straps.

Step 4   Using a power drill fitted with a ⁵⁄₆₄-inch drill bit, drill a hole in the base of the cannon trunnion mount. Then drill a hole at the top center of the large alarm clock bell.

Step 5   Grab a scrap of the 1/2-inch brass strip left over from making the vertical straps. We'll use this to create a sort of tracking mechanism for the proximity sensor, which directs the auto-cannon. Bend one edge of the track bracket 90 degrees. Then slightly round the corners.

Step 6   Mark, punch, and drill a hole to mount a miniature transit scope (which was once upon a time part of a key chain). The exact positioning of this mount will, of course, be entirely dependent on the pieces you have at hand and your choice of design. Please reference the photos with regard to your design process.

Step 7   Now we'll construct a sort of satellite dish type listening device/targeting antenna. This is designed to be part of the proximity sensor described in the fiction at the beginning of the chapter. You'll need a second small alarm clock bell from your box of bells (these tend to accumulate when you tear apart as many mechanical clocks as I do). Cut a 2 1/2-inch strip of 1/4-inch brass to fashion as a satellite mounting bracket. Use a 5/64-inch bit to drill a hole and attach the bell using an 8-32 × 1-inch brass machine screw, a 3/4-inch-long piece of 1/8-inch aluminum tubing, and a decorative threaded brass sphere (which I picked up in the lighting section of the local hardware store).

Step 8   Use a 5/64-inch drill bit to bore a hole in the other end of the mounting bracket, about 1/2 inch from the end. Attach the listening mount/targeting antenna to the track bracket (from step 5) using an 8-32, 1/4-inch brass machine screw and a #8 brass nut. It should look something like this when you're done:

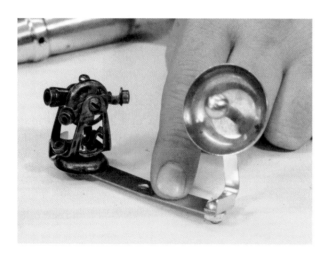

Step 9   If you can remove the top button from the pith helmet, now is the time to take it off. If your helmet is a cheap variation, like the one with which I am working, go ahead and drill a hole with a 5/64-inch drill bit directly into the top of the pith helmet.

Step 10   Assemble all the components on the top of the pith helmet. We'll use an 8-32, 1-inch brass machine screw inserted from the top down through the brass cannon trunnion, track bracket, a large spoked gear, and a reclaimed alarm clock bell. Secure it all in place (within the alarm bell) using a #8 brass nut. Then use another 8-32 brass

nut to secure it on the inside of the helmet through the hole drilled in step 9.

Step 11   I'll use four 6-32, 3/8-inch brass bolts and four 6-32 brass nuts to attach a couple of brass lion masks (commonly produced, or reproduced, for furniture restoration) on either side of the pith helmet near the ventilation holes, to give it a big-game safari feel.

Step 12   Assemble all of the components and attach the cannon barrel to the trunnion mount, as shown next.

Step 13   We'll add an old pinion gear from a cannibalized clock to the assembly, to drive the gear of the targeting assembly. Mask its base with a finish washer. You can see this addition in the photograph sort of peeking out from beneath the cannon barrel.

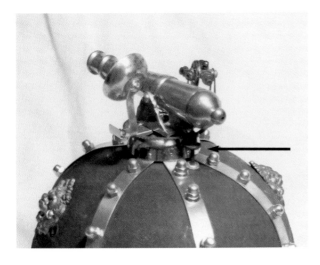

Step 14   To further give the impression that this is a fully functioning pith helmet we want to add a chin strap. Well no, not really, I don't want the chin strap to be usable, but rather would like to indicate its existence by adding a strap across the brim. Traditionally, these were made of leather, but for the example discussed here I have chosen to use a textured brass strap roughly 12 inches long. This sort of textured brass strapping is available through upholsterers and furniture restorers. I stretch the strap across the brim, fold it tighly around the edges, and secure it in place with a couple of rivets. Your strap addition will be a matter of design taste, and its method of attachment will be depended firstly on the construction of the helmet and secondly on whether or not you ever intend to wear the strap under your chin. With this final strap in place, your project is complete.

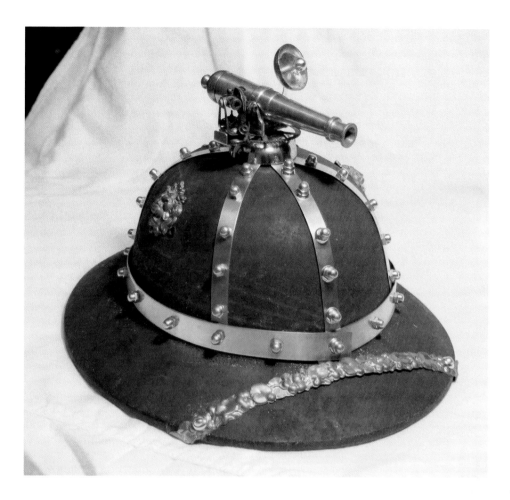

Note

As a fun little extra I found a small red light (which is supposed to be worn on your finger) and discovered that it fits nicely within the barrel of this cannon. So I have that option as well if I so choose.

Now that you have successfully completed your armored pith helmet and, as such, are suitably protected against even a swipe from a *Victoriasaurus regina*, might I suggest you try to adapt the principles to other sorts of wearable art? Perhaps try your hand at an armored top hat or bowler. Perhaps you need to augment a pair of boots. Anything is possible. Just be patient and avoid taking shortcuts. Build it properly and be worthy of the praise you will receive for your artistry and design.

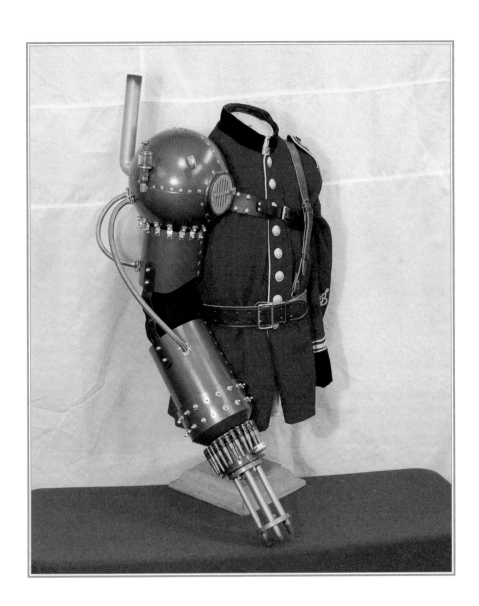

Chapter 12

Professor Grimmelore's Mark I Superior Replacement Arm with Integrated Gatling Gun Attachment

He cautiously knocked on the officer's door with his left hand, still unaccustomed to its use. A gruff voice bellowed from within, "Enter...."

He opened the door and stepped into the dark paneled room. "Ah, Lieutenant Harper. Do come in. No need to salute. I have read your commanding officer's report of the battle with the colonials. Nasty unfortunate business that. I do hope your recuperation is progressing well," said Colonel I.M. Havelleft, as he sat behind a large desk covered with disorganized papers.

"About as well as can be expected, I suppose, sir," replied Harper, looking down at the empty sleeve of his uniform hanging limply from the remains of his right shoulder.

The heavy-set colonel was also regarding the empty sleeve, accompanied by the sounds of whirring and clicking, as his mechanical eye focused in the dim light of the room. "Ah yes, about that..." he trailed off.

"Am I to be discharged, then, sir?" Harper asked regretfully.

The older man scratched his beard and surveyed the patriotic, if somewhat rash, young soldier carefully. "Well I suppose that is up to you, Lieutenant."

"I'm sorry, sir? I...I don't quite follow, sir."

"Certainly your wounds and sacrifice do indeed entitle you to an honorable discharge, but I see you are a man of conviction, and I would wager that there is a great deal of fight left in you, young man."

Harper thought before answering. "Yes sir...certainly sir.... But how would I...?"

"I am in a position to offer you a promotion, of sorts, as well as a transfer."

Harper did not reply, but his aspect brightened with a sort of confused anticipation.

"If you would consider a transfer to the B.O.R.G. 709th, you would be awarded the field rank of captain."

"The B.O.R.G., sir?"

"Indeed, the Britannic Officers Reconnaissance Group."

"Forgive me, sir, but I have never heard of the 709th, and I am uncertain how I could serve as a captain in my handicapped state. I mean, I cannot even hold a rifle."

The old colonel smiled, "Not to worry, lad. The 709th are the heavy mechanized augmented troops under the command of Major Montgomery. And I think you are a perfect candidate for their ranks." At this, a clicking, grinding pneumatic sound was heard as the colonel rose from behind his large desk.

Harper involuntarily took a half step backward as the old colonel rose to nearly 8 feet tall, supported from the waist down by a monstrous set of mechanical legs. Huge, heavy footsteps made the floorboards creak and groan as the colonel clomped his way ponderously toward the anteroom door.

"Come with me, Captain Harper. I think we can do something about that missing arm…."

—From *Resistance Is Bloody Futile: Adventures and Exploits in the Colonial Wars of the Britannic Officers Reconnaissance Group—709th*

My very first attempt at a mechanical arm back in 1996 was rather pathetic. It was made from copper sheeting and some old plumbing parts. Frankly, I am too embarrassed to show you a photograph of it. Nevertheless, the idea of a mechanical arm kind of haunted me until I built my first decent attempt in 2004. I had no real reason for building the arm other than to get the idea out of my head. The Steampunk community back then existed as little more than a few scattered like-minded lunatics in semi-regular contact through the Internet.

As it happens, I was to be merchanting corsetry at the October 2004 Whitby Gothic Weekend in North Yorkshire, England. I decided that my latest creation would be a wonderful accessory for the formal Goth ball, so I splashed out for the extra baggage allowance so I could bring it on the airplane with me and off I went, wondering if anyone would have any idea what to make of it. As it happens, very few did. But many were intrigued and wanted to know more.

I felt a bit like a missionary preaching the mad-science gospel or an ambassador from the land of Steampunk. Little could I know that in a few years' time, Steampunk would be a global aesthetic movement and be instantly recognized around the world. These days, when I attend Steampunk conventions and events (now that they actually exist), I am not surprised to see elaborate mechanical limbs.

Project Description

This chapter will detail the process for creating the Mark I type of steam-powered replacement arm that I initially designed in 2002. The project in this chapter is not an exact copy of that initial Mark I design, mind you; it has been improved and simplified for ease of construction. If you do manage to complete this project, please come show it to me at some event. After all, the B.O.R.G. 709th is always looking for new recruits.

As with many of the projects in this book, the ultimate design of your project is largely dictated by the parts and components that are available to you. Fortunately, in the case of this replacement arm project, most of these parts are available commercially from any decent hardware center.

I have already mentioned that two main axioms guide all of my construction and design: Metal is always preferable to plastic, and screwed is always better than glued. In designing and building this arm, we will be forced to abandon one of these philosophies and delve into the use of plastics. To make these components out of comparable metal parts would likely mean that the piece would be far too heavy to wear. The alternative, of course, is to start with thin metal, such as that used for ventilation or heating ducts. The problem with this approach, however, is that the thin metal does not look, sound, or move correctly, and it dents easily (not to mention that it tends to have very sharp edges, which could ultimately necessitate an actual replacement arm). For these reasons and others, which should become apparent as we proceed, we will be fabricating this project out of PVC plastic. Don't worry, it's going to look super cool when we're done.

What You'll Need

To complete the project as described in this chapter, you will need to get your hands on the following materials and tools.

Materials

- 10-inch-diameter "eyeball" plastic light cover
- 24-inch section of 6-inch PVC sewer pipe extension
- 4 × 6 inch PVC reducer coupling
- 10, 3/4 × 1/2 inch steel angle brackets
- 28, 10-24 × 1/2-inch steel machine screw
- 28, 10-24 acorn nuts
- 2 solid flanges for 1/2-inch pipe
- 5 feet of 1/2 inch PVC hot and cold feed line pipe
- 4-inch SDR plug gasket fitting
- 5, 6-32 × 1-inch brass machine screws
- 5, 6-32 brass nuts
- 32, #8 washers
- Scrap of leather, 3/4 inch wide by 5 inches long
- 1/2 inch × 5 inch long PVC nipple
- 1 1/2 inch × 15 inch waste arm (with coupling)
- 1 1/2 inch × 1 1/4 inch slip joint reducer coupling
- 32, 8-32 × 3/4-inch steel machine screws
- 32, 8-32 steel hex nuts

- 32, 8-32 steel acorn nuts
- 2 hinges from a Bell brand knee brace
- Sacrificial leather coat
- 4, #10 washers
- 20, 1/4-inch nickel finish Chicago screws
- Belt for chest straps
- 3-inch round drain grate
- Small, brass decorative hasp
- 1 oiler for a hit-and-miss steam engine
- 100 small double-cap nickel rivets
- 12 inches of 3/8-inch stainless steel faucet supply line
- 20 inches of 1/2-inch brass faucet supply line
- 4, 1/2-inch NM/SEU cable connectors
- Gauges and/or additional ornament

Tools

- Marker
- Safety glasses
- Power drill
- 3/4-inch spade bit or large boring bit
- Safety glasses
- Reciprocating saber saw or jigsaw
- Fine-toothed saw blade
- Metal file and/or sandpaper
- Masking tape
- 2 pairs of channel lock pliers
- Spring-loaded center punch or awl
- 3/16-inch drill bit
- Sandpaper or metal file
- Small sanding drum
- Rotary tool
- 5/64-inch drill bit
- Rotary hole punch

- Power saw or handsaw
- Saber saw or jigsaw
- Two-part epoxy or all-purpose cement
- Rust-colored spray primer
- Hammered silver spray paint
- Hammered gold spray paint
- Hammered dark bronze spray paint
- Flat-black spray paint
- Semi-gloss spray polyurethane
- Loctite or white glue
- Small adjustable wrench
- Scissors
- 13/64-inch drill bit
- Band saw, aviation snips, or rotary cutting tool
- 1/16-inch drill bit
- 1/8-inch drill bit
- Hacksaw or band saw with a metal-cutting blade
- 3/8-inch drill bit

Alternative Tools

- Anvil
- Hammer
- Vise

Stage 1: Start with the Shoulder

The primary piece for our arm is the part that will serve as the shoulder piece. For this, we will use a 10-inch-diameter "eyeball" plastic light cover.

It is best that you use a plastic light cover to replicate this design. Because plastic light covers are no longer fashionable, however, they can be tough to track down. Try lighting suppliers, and check to see if they can order one for you. Most of these are now supplied in glass (with the possible

exception of covers on college campuses, where students have a tendency to break the glass versions), but glass is not only difficult to work with, it is also challenging and dangerous to wear, so I certainly do not recommend it. In a pinch, a spherical plastic fishbowl, usually available from one of your large pet stores, will do.

The next piece you will need is a 24-inch section of 6-inch PVC sewer pipe extension.

Note

If this were the 1800s, this sewer pipe extension would have likely been made of either terracotta or cast iron, but you wouldn't want to try to wear either of those things on your arm as you would look quite silly.

Step 1   Position the plastic "eyeball" fixture on top of the sewer pipe extension (on end), such that the casting seam of the fixture is oriented like an equator, parallel to your work surface, and the opening (neck) of the fixture sits at a right angle to the sewer pipe and facing forward. Trace the pipe circumference onto the eyeball fixture. Mark this circle "B," for bottom.

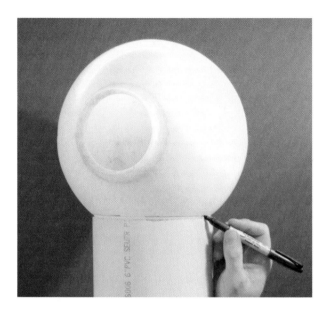

Step 2   Place a 4 × 6 inch PVC reducer coupling on your work surface, with the narrower opening down. Decide which arm, right or left, you intend to make.

Note

I assume your choices are limited to left or right, but if you have more than two arms, contact me directly for additional design specifications. Because I am left handed, I will be constructing a right arm.

Set the eyeball fixture on top of the reducer coupling, with the equator running vertically, and the opening (neck) at a right angle to the coupling. If you are making a right arm, your "B" circle should be on the left side (with the neck facing you). If making a left arm, the "B" circle should be on the right side. Trace the 6-inch circumference of the reducer coupling onto the fixture. Label the tracing either "R" or "L," depending on which arm you are making.

Note

From this point forward, I shall assume you are making a right arm that is identical to mine. You

should have learned enough through the various projects in this book to be able to adapt your patterns and designs as appropriate or necessary.

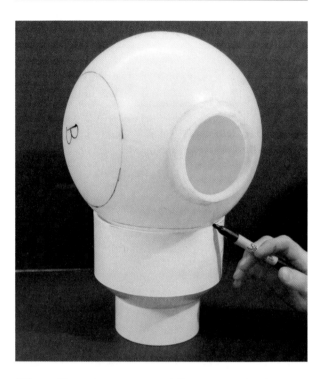

Step 3   Draw a second line, set 1/4 inch inside the circumference of circle "B." Connect this inner circle "B" tangentially with the circle you have labeled either "R" or "L," as shown in the illustration.

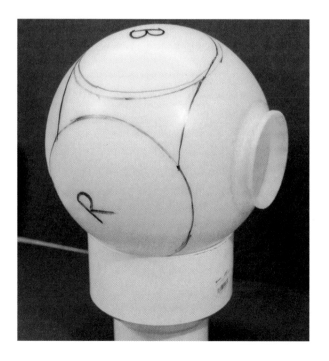

Hint

You may find it helpful to use your reducer coupling as a base for the "eyeball" fixture to help stabilize the pieces when drilling and cutting.

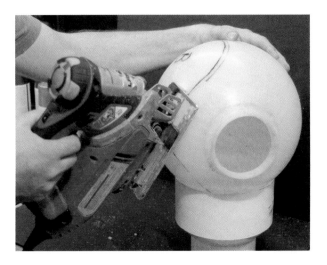

Step 4   Now we will cut out the hourglass shape we just drew. Wearing your safety glasses, use a power drill and a 3/4-inch spade bit or large boring bit to drill out a hole or two inside the hourglass shape.

When you are done cutting, the fixture should look like this:

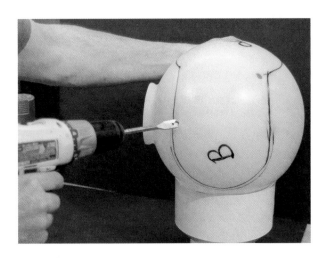

Note

For the sake of clarity, I have erased the extraneous marks from my "eyeball" fixture using some paint thinner. This is unnecessary as far as your project is concerned, but I thought it would make the process clearer.

After you have drilled out your insert holes for the saw blade, carefully proceed to cut out the hourglass shape using a reciprocating saber saw (or jigsaw) and a fine-toothed blade. Be certain to follow the manufacturer's safety instructions.

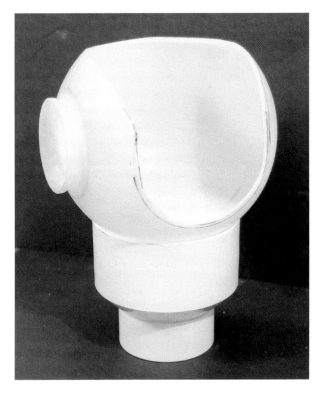

Step 5   Because no two people are built quite the same, try it on and assess whether it is digging into you at any point, front or back (most likely at the narrowest point of your hourglass). If so, trim away a bit more plastic. Continue this process until you are happy with how it covers your shoulder.

Note

When you try it on, be certain that the hole you had labeled "B" is positioned down by your arm, rather than up by your neck. The "eyeball" fixture opening should be positioned on the front of your shoulder.

Step 6 If you have followed along correctly thus far, you should now have a piece that somewhat resembles the helmets from the film *Spaceballs* (may the schwartz be with you). Next, you want to smooth off the internal edges where you cut. This part will be resting against your shoulder and triceps, so smooth it off carefully using whatever tools you have that seem most suitable—sandpaper, metal files, or even an electric rotary tool.

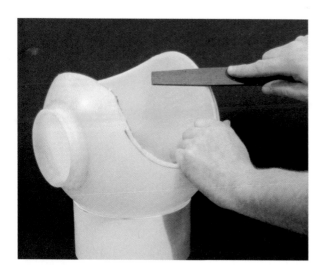

Stage 2: Create the Upper Arm

Step 1 Place the sewer pipe (try not to think about that…) on its end on your work surface, and position the cutout "eyeball" on top of it with the "B" circle facing down. On each side of the sewer

pipe, mark the two points at which the cutout intersects the pipe.

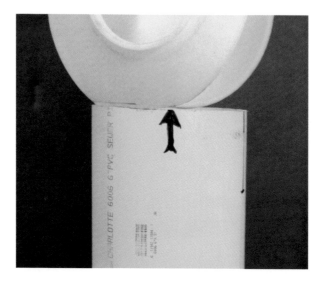

Step 2 Halfway between the two marks you just made, measure down 3 inches and make a mark. Scribe an arc that connects the two marks you made in step 1 and passes through this 3-inch mark, as shown:

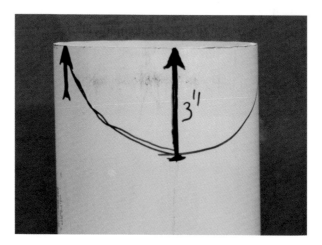

Step 3 Using a jigsaw or reciprocating saber saw, cut out this half circle from the sewer pipe. Smooth out the edges with a file or sandpaper. The following image shows how the "eyeball" now matches up with the cutout sewer pipe.

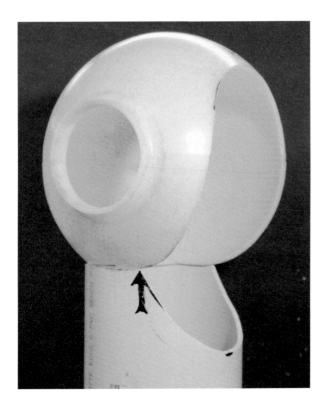

Step 4 Measure down 6 inches from the top edge of the PVC pipe, and draw a line around the pipe. Then cut along this line. If you're using a saber saw, drill a hole first, and then run the saw blade through the hole. Otherwise, you can just saw through the pipe with a handsaw of any type.

Note

This piece of PVC pipe from step 4 will serve as the upper arm (biceps/triceps) part. If you are creating this project for someone with exceptionally long or short arms, this 6-inch length (which seems to be a good average) might need to be adjusted longer or shorter, as necessary.

Step 5 Sand or file the cut edge.

Step 6 Using some masking tape or duct tape, attach the shoulder (the "eyeball" fixture) temporarily to the upper arm/bicep section (oriented as shown in the adjacent illustration). At the base of the upper arm pipe, mark a point that is directly below the "eyeball" fixture opening (neck). Measure up 1 1/2 inches from this center mark and mark another point. Then measure and mark 3 inches to either side of the center mark you made at the base of the upper arm pipe, and scribe an arc from one side to the other, passing through the 1 1/2-inch mark, as shown:

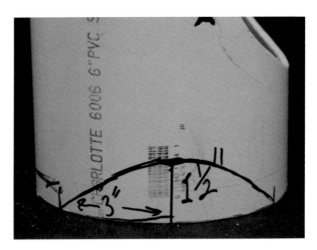

This arc will accommodate the wearer's inner elbow (so that the wearer can actually bend his or her arm). Go ahead and cut out that arc, and finish off the edges. Take apart the two pieces if you need to before making the cut.

Stage 3: Attach the Upper Arm to the Shoulder Boiler

Because we will be bolting these various parts together, we can attach the *shoulder boiler* (formerly "eyeball" fixture) to the *upper arm section* (formerly the sewer pipe/bicep extension).

Step 1   If you took apart these two pieces to cut out the inner elbow, go ahead and tape them back together now.

Step 2   Although you could probably attach these two pieces together permanently in several different ways, we are going to use ten, 3/4 × 1/2 inch steel angle brackets. The angle of the seam where our upper arm meets our shoulder boiler is wider than a right angle. In fact, it is closer to a 130-degree angle. We want, therefore, to open up the angle brackets to a greater angle. You can do this using two pairs of channel lock pliers.

Alternatively, you could set the bracket (angle up) on an anvil and hammer it open to the desired angle. Or you could place the bracket angle in a

vise and crank the jaws tighter until the bracket opens up to the angle you want. (Hmmm…that sounds like a sentence out of *The Big Book of Pain and Torture*.) Whichever method you choose, make sure that each bracket angle matches up closely with the angle formed at the junction of the upper arm piece and the shoulder boiler. (Refer to the illustration after step 6 to see how these will be attached.)

Step 3   Draw a vertical line from the bottom center of the manufacturer's circular opening in the shoulder boiler, down to the top center of the inner elbow cutout. Position your first bracket on this line at the seam, and mark the position of the bracket's holes on the shoulder boiler and the elbow cutout.

Step 4   Proceed systematically around the seam, marking the positions of the holes for all ten brackets, ensuring that they are spaced apart evenly. Place one at each end of the arc, then place the remaining eight around the arc, roughly evenly spaced.

Step 5 Use a spring-loaded center punch to mark the places where you will drill, to minimize the chance of your drill bit sliding off to the side. Then drill out the marked holes using a 3/16-inch bit fitted into a handheld power drill.

Step 6 Pull off the tape and disassemble the two pieces. We'll reattach the two pieces using the steel angle brackets. Slot ten, 10-24 × 1/2-inch steel machine screws through the drilled holes in the top of the upper arm from the inside out. Position a bracket in place and secure with a 10-24 steel acorn nut.

Step 7 Place the shoulder boiler in position on top of the upper arm. Align the holes with the brackets and secure in place using ten more 10-24 × 1/2-inch steel machine screws and ten 10-24 steel acorn nuts.

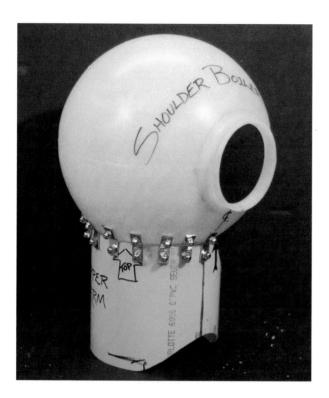

Stage 4: Build the Gatling Fist

Let us switch our attention to the other end of the arm and work our way back from the hand to the elbow.

Step 1 From the plumbing section of your local hardware center, select two solid flanges for 1/2-inch pipe, shown next. Also acquire a 5-foot length of 1/2-inch PVC hot and cold feed line pipe.

Step 2 Remove all labels and adhesive residue. (Goof Off brand solvent works exceptionally well for this purpose.)

Step 3 Check to see if the 1/2-inch PVC pipe will fit into the perimeter holes on the flanges. If the holes are too small (mine were), open them up a bit using a small sanding drum on a rotary tool.

Step 4 On one of the two flanges, make a mark between each of the 1/2-inch perimeter holes (four marks total). Drill out holes at these marks using a 5/64-inch drill bit fitted into a power drill.

Step 5 Grab a 4-inch SDR plug gasket fitting (again from the plumbing section), as shown next. On top of it, where it flares out, center the flange you just drilled out. Use an awl or center punch to mark the position of the holes you drilled onto the cap of the plug gasket.

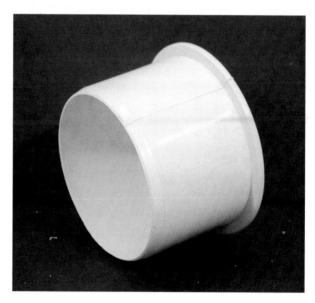

Hint
You might find it useful to make alignment marks on both the flange and the plug gasket so that you can realign the position of the holes.

Step 6 Drill out the holes in the plug gasket using the 5/64-inch drill bit.

Step 7   Slot four, 6-32 × 1-inch brass machine screws through #8 washers and then, from the inside of the plug gasket, slide two opposite screws through the holes you drilled in the flange. Secure these in place using 6-32 brass nuts.

Step 8   Now we'll add a leather strap that will provide a bit of a hand hold on the inside of the Gatling fist. This will help to support the weight of the entire arm while you are wearing it. Cut a scrap of leather roughly 3/4 inch wide by 5 inches long, and punch holes 1/4 inch from each end. Slot the

two remaining machine screws (with washers) through the leather strap on the inside of the gasket before slotting them through the flange.

Henceforth we shall refer to this assembly as *the fist*.

Step 9   Screw a 1/2-inch × 5-inch-long PVC nipple into the center hole of the flange part of the fist. Then screw the other flange onto the other end of the PVC nipple. Line up the 1/2-inch holes around the perimeter.

Step 10   Measure and cut (almost any type of saw will work) four, 7 1/2-inch lengths of 1/2-inch hot and cold PVC feed line pipe. Clean up the ends and edges of your cuts with sandpaper if necessary.

Note

Do not simply measure the PVC pipe in 7 1/2-inch intervals and cut, however, because the saw blade width will make them slightly off, and they need to be exact. Instead, either set a guide fence for your power saw or cut one piece at exactly 7 1/2 inches, and then use it to mark the others. Each 7 1/2-inch length of pipe must be exactly the same length.

Step 11   Slot the barrels you have just cut into the outer holes on the flange and fist, and marvel at what thou hast wrought. *Do not,* however, fix or adhere any of these in place at this stage. They will need to be removed before you paint your arm.

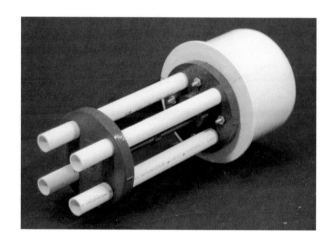

Step 12   Take the entire Gatling fist assembly and insert it into the 4 × 6 inch PVC reducer coupling. Mark a line where the plug gasket part of the fist meets the top edge of the PVC reducer coupling, as shown next. Use masking tape to mask off below this line (where it will be inside the reducer coupling, at the wrist). You do not want to paint below this line.

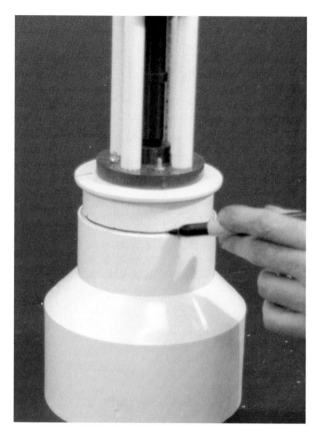

Stage 5: Add the Forearm

Step 1   Measure and cut another section of 6-inch PVC sewer pipe extension to a length of 9 1/2 inches. As with the upper arm section, you might need to adjust this length for significantly long or short arms.

Step 2   Smooth off the cut edges using sandpaper, metal files, or an electric rotary tool.

Step 3   Insert the forearm section into the reducer coupling wrist. Ensure that it is fully and firmly seated.

Step 4   Mark out a double row of holes that are evenly spaced and approximately 1 1/2 inches apart and set in about 3/4 inch from the top and bottom edges of the widest point of the wrist coupling, as shown next.

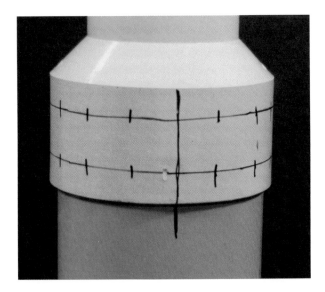

Step 5   Use an awl or spring-loaded center punch to preset the holes, and then drill them out using a power drill fitted with a 3/16-inch drill bit. You should drill through the wrist coupling as well as the PVC pipe.

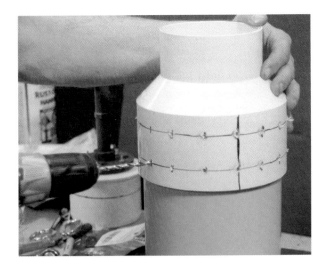

Stage 6: Make the Boiler Stack

Step 1   For the boiler stack for your internal shoulder furnace, you'll need a 1 1/2 inch × 15 inch waste arm (with coupling) and a 1 1/2 inch × 1 1/4 inch slip joint reducer coupling, as shown next.

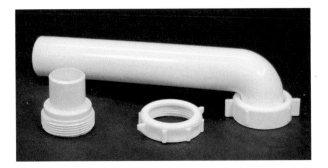

As with almost all of the components used in this project, you'll find these in the plumbing section.

Step 2   Cut down the waste arm to the desired length. In this example, I cut off about 6 inches, leaving about 8 inches.

Step 3   Inside your shoulder boiler, mark a spot on the equator line directly opposite the front opening. I have drawn in the equator line for reference.

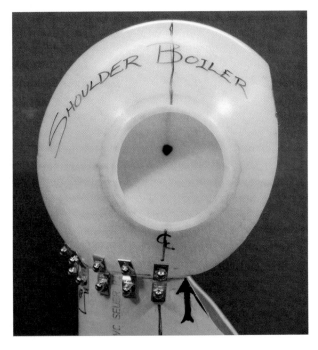

Step 4   Trace the external circumference of the larger part of the slip joint reducer coupling.

Then use a drill and/or saw to cut out the hole. It is important that you cut out this hole a tiny bit

smaller than the circumference you traced, and then slowly enlarge the hole using a file or a drum sander, as needed, as you work on fitting in the slip joint reducer coupling into the hole. You want this to be a nice, tight fit.

Step 5 Thread the slip joint reducer coupling into the hole you have just cut or drilled, and use a two-part epoxy or all-purpose cement to glue the reducer coupling in place, like so:

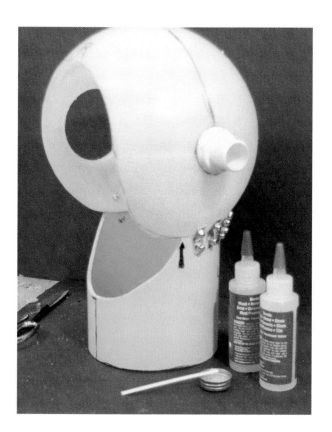

Stage 7: Add Primer to Your Arm

Now we will disassemble all the component parts that you have assembled thus far and prime them. I have chosen a rust-colored spray primer. On occasion, even despite the thorough polyurethane clear-coating I applied over a piece, chips of paint can come off. When this happens, the rust-colored undercoat shows through, further enhancing the impression of the piece actually being cast iron (or whatever).

When you have finished coating all of your plastic/PVC parts with spray primer, allow them to dry thoroughly overnight.

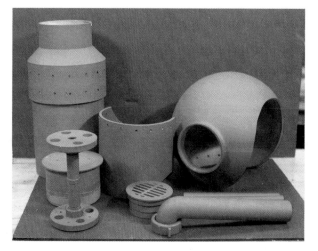

Note

This image shows the 3-inch drain grate we'll add later. You'll need to prime this, too, so you might as well add it to the paint pile now.

Stage 8: Paint Your Arm

Having thoroughly primed all of the plastic parts, we can proceed to paint the various parts and components with different metallic colors and tones of spray paint.

For my arm, I will paint my gun barrels and the 3-inch round drain grate and the smoke stack (all of which we will visit a bit later) with a hammered silver spray paint. I will paint the Gatling frame assembly with a hammered gold spray paint, and the large pieces (upper arm, shoulder boiler, forearm, and wrist/fist) will be covered in a hammered dark bronze spray paint. I use just a little bit of flat-black spray paint to "char" the ends of the Gatling barrels and the tip of the smoke stack. This gives them a nicely used look.

The instructions on the side of the cans of spray paint recommend that you allow it to dry and harden for 48 hours. Who am I to argue? So I advise that you wait 48 hours before moving on.

Hint

While you are waiting for these components to dry, you could probably start and complete one of the other projects in this book. Just thought I would mention that.

After the paint is dry, proceed to clear-coat all of your painted components. I recommend using a semi-gloss clear spray polyurethane. Read and follow all instructions on the spray can. Apply at least two even coats of the polyurethane spray. Wait the recommended 24 hours for it to cure, dry, and harden completely. Remove any masking that you applied to your pieces prior to priming.

Stage 9: Reassemble the Hand and Gatling Fist

Step 1   Repeat the assembly steps (5–7) of Stage 4 to attach the Gatling device to the plug gasket fist.

Step 2   After you have bolted all the parts in place (remember to include the leather strap on the inside), you will proceed to slide the Gatling barrels into place. As you slot the barrels into place, use some two-part epoxy to secure them firmly.

Step 3   Set aside the entire Gatling fist assembly and wait for the glue to dry.

Stage 10: Reassemble the Forearm and Wrist

Step 1   After the glue on the forearm and wrist pieces is thoroughly dry, proceed to fill each of the holes you drilled earlier in Stage 5, step 5, using 8-32 × 3/4-inch machine screws and #8 washers. Then secure them in place with 8-32 hex nuts and 8-32 steel acorn nuts.

Step 2   Use a dab of Loctite or white glue to secure the nuts and acorns onto the ends of the machine screws, to mitigate the possibility of them coming loose through vibration or wear.

Step 3   Use a small adjustable wrench to tighten the nuts in place, but be careful not to over-tighten them lest you risk cracking your beautifully painted PVC construction.

Stage 11: Reattach the Upper Arm to the Shoulder Boiler

Step 1   Reassemble the upper arm piece and the shoulder boiler, as you did through Stage 3, step 6, using all the hardware you removed in Stage 7.

Step 2   Use a dab of Loctite or white glue to secure the nuts and acorns onto the ends of the machine screws.

Step 3   Wrench-tighten the nuts in place, but be careful not to over-tighten them lest you risk cracking the PVC. After you have reassembled these pieces, you should have three distinct arm features, as shown below.

Stage 12: Create the Elbow Hinge

Step 1   You now need to find something to use as an elbow hinge. Any hinge that will open from 90 to 180 degrees on a lateral plane (as opposed to most hinges, which operate on a vertical plane) should serve for your purpose. Friends have employed objects such as nut crackers or can openers for this purpose, but you could easily make this hinge yourself.

I am using two hinges from a Bell brand knee brace, which is available from most pharmacies or sporting goods stores. I have chosen this particular type of hinge because, firstly, it is specifically designed to mimic the movement of simple pivot joints such as knees and elbows; and, secondly, it is a three-part hinge (as shown next), which allows a more fluid range of motion than can be experienced with just a simple pivot hinge.

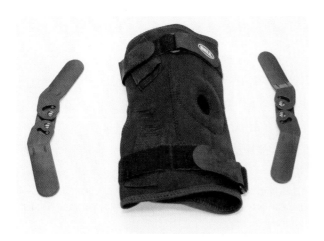

Note
I will proceed under the assumption that you are using the same sort of hinges that I am using.

Step 2   Mark where you will drill rivet holes on each end of each hinge. These holes should be centered across the width of the hinge, with the first hole 3/4 inch from the end and the second hole set in 2 inches from the end.

Step 3   Use an awl or spring-loaded center punch to dimple and preset the holes you are about to drill, and then drill out the holes using a 3/16-inch drill bit.

Step 4   On the inside of one of the elbow hinges, mark an "I" for inside; on the inside of the other hinge, mark "O" for outside.

Step 5   Position the outside hinge ("O") off the bottom edge of the upper arm section at a point 90 degrees around the circumference of the arm from the front (cut out) opening. (See the next illustration for reference.) Mark the locations of the holes onto the upper arm piece.

Note
The lower hole is 3/4 inch from the bottom edge of the upper arm piece. Unless this replacement arm is being made for a mutant, you will want to ensure that the bend of the hinge is to the front (which is, of course, why we have designated the hinges inside and outside).

Step 6 Preset the holes on your upper arm piece, and drill them out using a 3/16-inch drill bit.

Step 7 Repeat this process for the inner part of the arm, directly centered under the cutouts of the shoulder boiler (see illustration). Note that on this side, however, the hinge will need to bend in the opposite direction.

• **Step 8** On the inner side of the upper arm section, preset (dimple) and drill out the holes using a 3/16-inch drill bit.

Step 9 Bolt the hinges onto the upper arm section using four, 10-24 × 1/2-inch steel machine screws and 10-24 steel acorn nuts.

Step 10 Mark a hole position 3/4 inch from the edge of the forearm piece, with a second hole set in 1 1/2 inches from the first.

Step 11 Preset (dimple) and drill out these two holes using a 3/16-inch drill bit.

Step 12 Do the same where the other side of the hinge will attach to the upper arm section.

Step 13 Bolt the hinges onto the forearm section using 10-24 × 1/2-inch steel machine screws and 10-24 steel acorn nuts.

Note

Do not use Loctite or glue to secure these nuts and bolts if you intend to, or think you might, transport the arm in the future. The ability to unbolt one end of the elbow hinges makes for far easier packing.

Step 14 Hooray. Your random assortment of plumbing parts is starting to resemble a mechanical arm. Go ahead. Put it on. Flex and pose a bit. You know you want to. When you have finished patting yourself on the back, proceed to Stage 13.

Stage 13: Add the Edging and Lining

If you have great sewing skills (or know someone who does), you can make a sturdy suede or leather sleeve to wear inside your mechanical arm. This will protect your actual arm from the mechanical contraption you are constructing and will also help hide your actual biologicals. If you do not want to sew up a bespoke sleeve, you can do as I have done in this project (and most of the projects in this book): scavenge.

Step 1   Find yourself an old, secondhand leather or suede coat or jacket. I found a suitable candidate at a charity shop just down the road for about $8. Oh…and make sure it fits.

Note

A useful alternative to cutting up an old coat is to use welding sleeves (available from most places where welding supplies are sold), but they can be more expensive and can be tougher to find. It is entirely your choice.

Step 2   Cut off a sleeve of your sacrificial coat and remove any lining. If your mechanical arm will be your right arm, cut off the coat's right sleeve; if it's the left arm, cut off the left sleeve.

Step 3   Cut off the coat's zipper guard or button guard, if it has one, or cut a 2 1/2 × 22-inch strip of leather from the coat. You should have two pieces that look like what's shown at the top of the next column.

Step 4   Near the shoulder seam of the sleeve on the inside, mark four holes that match up with

the four middle bolts of the angle brackets on the shoulder boiler. Punch out these holes using the third-largest setting on a rotary hole punch. This is another one of those cases in which it is far easier to show you what I mean than to try to find the right way of describing it. To that end, please examine the following two photos (on this page and the next) until it makes sense.

Step 5   Remove those four bolts from the shoulder boiler by undoing the acorn nuts on the outside. Then add four #10 washers on the leather side before replacing the bolts, through the leather and the shoulder boiler, and again securing them in place with the steel acorn nuts.

Step 6 Now we will create a leather gasket to run around the inside of the shoulder boiler cutout area. This will be the gasket that sits on top of your shoulder by the base of your neck and extends around front and back toward your armpit. Beginning about 1/2 inch from one of the armpit edges, mark a hole position 1/4 inch from the edge of the boiler cutout area. Mark more hole positions at 1 1/2 inch increments.

Step 7 Use an awl or spring-loaded center punch to preset or dimple these holes, and then drill them out using a 13/64-inch drill bit.

Step 8 Take that strip of leather (in my case the former zipper guard) you cut in step 3, and lay it around the edge you just marked and drilled,

ensuring that the finished edge of the leather strip (if it has one) will be toward the outside.

Step 9 Mark the positions of two rows of holes onto the leather gasket strip based on the holes you made in the shoulder boiler in step 7, about 1/2 inch from either long edge and at intervals along its length.

Step 10 Using a rotary hole punch set on the second-largest hole setting, punch out both rows of holes in the leather. This process is very similar to that described in Chapter 5 (Stage 6, steps 8–12) on making goggles—although this gasket is obviously somewhat larger.

Step 11 Wrap the leather gasket strip around the edge of the shoulder boiler cutout ("R") and secure it in place with 1/4-inch Chicago screws (available from most leather supply shops or some craft centers—and, of course, online). You might want to Loctite or glue the Chicago screws to hold them in place.

Note

If you do not have or cannot find Chicago screws, you could also use medium double-cap rivets or even 3/8-inch machine screws and nuts.

When you are finished, your shoulder boiler should look something like this:

Stage 14: Add Straps

Step 1   Find an old belt (charity shops are useful for this) that is plenty long enough to reach around your chest, just below your armpits.

Step 2   Mark a point on the back side of the shoulder boiler, about 3/4 inch below the equator

line and equidistant between the edge of the gasket and the attachment fitting for the smoke stack.

Step 3   Mark a corresponding point on the front side of the shoulder boiler.

Step 4   Drill out both of these holes using a 13/64-inch drill bit.

Step 5   Put on the shoulder. Hold the buckle of the belt over your heart and run it under the left arm (assuming you are building a right arm) and across your back to the hole you marked and drilled on the back of the shoulder boiler. Cut off the belt 3/4 inch beyond this hole and clip off the corners for finish.

Step 6   You should now have two sections of belt. The long end with the buckle will attach to the back of the shoulder boiler near your shoulder blade. The shorter tongue end of the buckle will attach to the front of the shoulder boiler near your armpit. Punch holes 3/4 inch from the cut-off ends of these straps and secure them in place on the shoulder boiler using 1/2-inch Chicago screws.

Note

Be certain to Loctite or glue the Chicago screws to keep them together.

Further Note

Chicago screws are best for this purpose, because you will want some play in the attachment point so that the angle of the belt is versatile and variable. The belt should be able to swivel a bit on the Chicago screw.

Stage 15: Add Details

We are now ready to turn our attention to a variety of little details that will serve to make this arm something amazing. After all, what is the point of you spending all this time building this if your friends (and nemeses) aren't going to be absolutely awed by it?

At this stage, you might want to remove the chest strap that you just attached, to keep it out of the way while you work on other decorative elements.

Detail A: The Boiler Door

Step 1   For the boiler door, you will need a 3-inch round drain grate. Paint it silver (if you haven't already done so).

Step 2   After the paint is dry, mix up some two-part epoxy.

Step 3   Use the two-part epoxy to glue the grate into the opening at the front of the shoulder boiler.

Note

By luck or happenstance, the opening on the original "eyeball" fixture that serves as the body for this shoulder boiler is exactly 3 inches in diameter, which made the drain grate a perfect marriage. If yours is not exactly the right size, you might have to adjust the hole to accommodate the boiler door.

Step 4   Take a small, brass decorative hasp and cut off the loop slot using a band saw, aviation snips, or rotary cutting tool.

Step 5 Using a 1/16-inch drill bit, enlarge the holes in the bale loop, if necessary.

Step 6 Mark the positions of the holes of the bale loop onto one side of the boiler door. The loop is going to serve as a handle. (Well, actually, the door is glued shut, so it is only pretending to be a handle.)

Step 7 Mark the positions of the holes in the hasp hinge on the other side of the door.

Step 8 Drill 1/16-inch holes about 1/2 inch deep in both of these locations.

Step 9 Use two-part epoxy or all-purpose cement to attach the loop and the hasp hinge. Then glue small brass brad nails (which usually come with the hardware at the time of purchase) into the holes you just drilled.

Assuming you have followed along with each of these steps in the same way that I have described, your boiler door should now look like this:

Hint

You should feel free to be imaginative with this project. Consider inserting a piece of red cellophane or Plexiglas on the inside of the door, perhaps with one of those small flickering battery-powered tea lights glowing behind it. For you more serious makers, suppliers of model train set components offer LED boiler lights you can use. Of course, if you are savvy enough to know where to get and how to install such things, you probably don't need me to explain it to you.

Detail B: The Oiler

Step 1 Get yourself an oiler for a hit-and-miss steam engine. The one I am using was acquired at a local flea market, but these are easily available online.

Step 2 Make a mark along the shoulder boiler equator, which is directly above where the outer elbow hinge will attach—this mark will be at the outermost point of the external curve. This is where you will be attaching the oiler coupling (see the image with step 4).

Step 3 Preset and drill out a hole using a 3/8-inch drill bit.

Step 4 Thread the oiler coupling into the hole you just drilled. Fix it in place using some two-part epoxy or all-purpose cement before securing it with a corresponding nut on the inside. (The size of the hole, nut, and coupling will, of course, be entirely dependent on the size and type of the oiler you use.)

Step 5 Use a dab of two-part epoxy or all-purpose cement and push the posts for some small double-cap nickel rivets into these holes.

Step 6 Continue to create faux seams and rivet rows until you are content with the design.

Hint

If you do not have and cannot find an oiler, consider using a gauge of some type. If you do not have or cannot find a suitable gauge, consider making one for the purpose based on the instructions in Chapter 6.

Detail C: Faux Rivet Seams

Step 1 Scribe a line from front to back, equivalent to the prime meridian relative to the equator seam (that is, is should be perpendicular to the equator).

Step 2 On one side of the equator seam, mark points at 1-inch intervals. Do the same for the prime meridian line, as well as points around the boiler door.

Step 3 Optionally, you can continue scribing these lines down the length of the upper arm and then continue again on the forearm.

Step 4 Preset and drill out all of these holes using a 1/8-inch drill bit.

Detail D: Cable Pipes

One of my favorite types of decorative detail on a mechanical arm is the addition of faucet supply lines, or cable pipes; they lend the assembly a sort of pneumatic quality.

Step 1 Purchase two faucet supply lines from your local hardware center or plumbing supplier. They come in a variety of lengths and diameters as well as both stainless steel and brass finishes. For this project, we will use 12 inches of 3/8-inch stainless steel faucet supply line and 20 inches of 1/2-inch brass faucet supply line.

Step 2   Because I do not have the correct sort of coupling to thread into the ends of these cable pipes (as they will henceforth be deemed), I begin by cutting off both ends of both cable pipes, as shown, with standard hacksaw, rotary cutting tool with a cut-off wheel, or band saw with a metal-cutting blade.

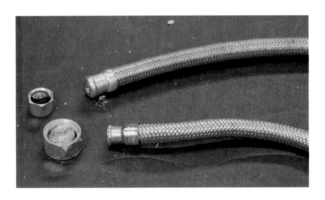

Step 3   Now you will need four, 1/2-inch NM/SEU cable connectors from the electrical supply section of the hardware store.

Step 4   Determine the positioning of your first cable pipe, and mark the locations of the ends. In this case, we will position the top end of the short stainless steel cable pipe to the shoulder boiler about 1 1/2 inches below the equator line and about 2 inches from the smokestack vent. The lower end of this cable will attach down near the bottom of the triceps on the upper arm section.

Step 5   Drill out these marked points using a 3/4-inch spade drill bit fitted into your power drill.

Hint

In case you are worried about messing up your project, it might be worth practicing drilling these holes or even checking the fit of your cable connector coupling by trying them out on a scrap of cut-off sewer pipe left over from Stages 2 and 3.

Step 6   Remove the cable connector collar and thread the connector into the holes you just drilled. If your holes are the right size and drilled out cleanly, the metal thread on the connectors should simply cut its way through the PVC pipe. If they are slightly loose, you might need to help secure them in place with both the connector collar and a bit of epoxy or all-purpose cement.

Step 7   Clamp the ends of the cable pipe into the cable connectors you just installed. In the case of this upper arm cable pipe, because I am unlikely

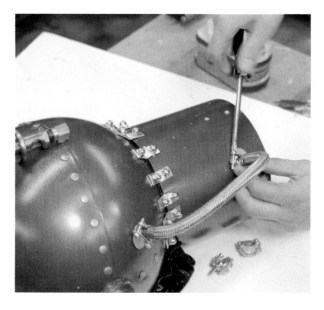

to need to remove this on a regular basis, I will Loctite or glue the cable connector clamp screws in place.

Step 8   Now determine the locations for your second cable pipe—the 20-inch brass cable pipe.

Note

Although the cable pipes might be different sizes, the diameter of the cable connectors is identical.

Position the top end about 2 1/2 inches forward of the top end of the stainless steel cable pipe. Locate the bottom end about 2 1/2 inches inside the lower end of the outer elbow hinge.

Step 9   Drill out both of these holes, as you did in step 5, using your 3/4-inch spade bit.

Step 10   Repeat step 6, ensuring that the cable connectors are firm and secure.

Step 11   If you now want to insert the cable pipe and clamp it all together to check for fit, by all means, go ahead. Note, however, that to complete the final stages, it will be easier to work on the shoulder boiler if it is separate from the forearm.

Detail E: Gauges

I have decided to add a couple of handmade gauges (see Chapter 6) to this arm. You could, of course, buy a new or antique gauge for this purpose, but it is far more gratifying to use gauges that you have constructed yourself.

Step 1   Make or acquire a gauge.

Step 2   Determine your location for the gauge. I have chosen a point just above and outside of the boiler door, which will still permit the wearer to read it.

Step 3   Mark and drill a hole at that location using the 5/64-inch drill bit.

Step 4   Assuming that you are using one of the gauges from Chapter 6 that you have made yourself, you should see an exposed 6-32 machine screw thread, which easily slots through the hole. Secure it in place using a washer and standard 6-32 nut.

Note

If you are using a gauge that you purchased, you will have to devise your own means of attaching it to your mechanical arm.

Step 5   Repeat steps 1 through 4 as many times as you wish to add as many gauges of as many sizes and as many types as you desire.

Detail F: Smoke Stack

Step 1   Mix up some two-part epoxy and adhere the smoke stack firmly to the smoke stack vent.

Note

It would be wise to sand or file away the paint a bit so that the epoxy can get a firm grip, plastic to plastic, rather than plastic to paint.

Step 2   Screw the threaded collar securely on to the corresponding smoke stack vent thread.

Step 3   Wait for the epoxy to set and dry.

Detail G: Other Stuff

We have finished adding details and decoration for this project, but, of course, you can continue to add more if you wish. Take it is far as you dare. You still have a lot of ornamented space to work with if you want to, so go ahead and unleash your creativity. I added a chain of plastic bullets I found at the local costume shop; I think they really help with the Gatling effect. I also added another gauge. I cannot wait to see what sorts of variations people come up with.

Stage 16: Assemble the Arm

Step 1   Reattach the elbow joints connecting the upper arm and forearm. Secure these in place with acorn nuts.

Note

If you ever intend to travel with this arm, you would be wise to avoid using Loctite or glue to affix these four acorn nuts in place. It is much easier to take the two halves apart this way.

Step 2   Run the leather sleeve down through the assembly.

Step 3   Connect or reconnect the long brass cable pipe.

Step 4   Reattach the cross chest buckle strap both front and back.

Step 5   Insert the Gatling fist into the wrist.

Step 6   Try it on for size, fit, and intimidation factor.

Step 7   Amaze your friends and harass your enemies.

Part the Third

Hastily Scribbled Laboratory Notes

Appendix A

This Way Lies Madness

$\mathcal{T}$ have to operate under the assumption that if you are reading this book, you are at least somewhat familiar with the whole genre of Steampunk. But in my experience, even Steampunk experts frequently fall short when it comes to the practicalities of design. To that end, I wanted to include here a list of recommended reading and/or viewing to help you get a better sense of the fantastical aesthetic, but I quickly came to realize that to do so would require an entire book unto itself.

I have decided to include, therefore, only a very, very select list of suggested readings and viewings. This is not to indicate in any way that these are the very best that the genre has to offer, but rather that these titles include design elements or inspirations that have been formative in my own development as an artist, designer, and mad scientist. I hope you find them all as useful and as enjoyable as I have. I've thrown in a few obscure references you are welcome to toss about to impress your friends and baffle your foes.

Further Reading

Appleton, Victor. 1910–2007. *Tom Swift Series*. Various publishers.

Foglio, Phil and Kaja. 2001–2011. *Girl Genius*, comic series. Seattle: Studio Foglio.

Gibson, William, and Bruce Sterling. 1991. *The Difference Engine*. New York: Bantam Spectra.

Guinan, Paul, and Anina Bennett. 2000. *Boilerplate: History's Mechanical Marvel*. New York: Abrams Image.

Leopold, Allison Kyle. 1986. *Victorian Splendor: Re-creating America's 19th-Century Interiors*. New York: Stewart Tabori & Chang.

Moore, Alan. 1999–2007. *The League of Extraordinary Gentlemen*, comic series. La Jolla, California: America's Best Comics.

Nevins, Jess. 2005. *The Encyclopedia of Fantastic Victoriana*. Austin, Texas: MonkeyBrain Books.

Pullman, Philip. 1995. *The Golden Compass*. New York: Random House.

Ruhling, Nancy, and John Crosby Freeman. 1994. T*he Illustrated Encyclopedia of Victoriana: A Comprehensive Guide to the Designs, Customs, and Inventions of the Victorian Era*. Philadelphia: Running Press.

Sears Roebuck & Co. 2007. *1897 Sears Roebuck & Co. Catalog*. New York: Skyhorse Publishing.

Stephenson, Neal. 1995. *The Diamond Age*. New York: Bantam Spectra.

Although you really should read books by H.G. Wells, Jules Verne, and Mark Twain, you should also keep a sharp eye on a few contemporary authors (in alphabetical order):

- Pip Ballantine
- Anina Bennett

- James Blaylock
- Greg Broadmore
- Mark P. Donnelly
- G.D. Falksen
- Phil & Kaja Foglio
- William Gibson
- George Chetwynd Griffith
- Paul Guinan
- Jay Lake
- Tee Morris
- Tim Powers
- Cherie Priest
- Neal Stephenson
- Bruce Sterling

Viewing

The Adventures of Brisco County, Jr. (TV series; Carlton Cuse, Jefrey Boam, 27 ep. 1993–1994)

Back to the Future, Part III (1990)

Chitty Chitty Bang Bang (1968)

City of Lost Children (1995)

The Golden Compass (2007)

The Great Race (1965)

The League of Extraordinary Gentlemen (2003)

Those Magnificent Men in Their Flying Machines (1965)

Oblivion & Oblivion 2: Backlash (1996)

Perfect Creature (2006)

Rayguns Promotional Transmission (commercial: www.wetanz.com/rayguns)

Robot Carnival (animation, 1991)

Sherlock Holmes (2010)

Steamboy (animation, 2004)

Time After Time (1979)

The Wild Wild West (TV series, 1965–1969)

Wild Wild West (feature movie, 1999)

Gaming

Chadwick, Frank. 1988. *Space: 1889* RPG series. Heliograph, Inc., Untreed Reads Publishing LLC.

Hensely, Shane Lacy. 1996. *Deadlands: The Weird West Roleplaying Game* series. Pinnacle Entertainment Group.

Pondsmith, Mike. 1994. *Castle Falkenstein* RPG series. R. Talsorian Games, Steve Jackson Games.

Stoddard, William. 2000. *GURPS Steampunk* RPG. Steve Jackson Games.

Witt, Sam. 1993. *Wooden Suits & Iron Men* RPG. Nightshift Games.

Appendix B

Dramatis Personae

Here is a list of characters who may or may not exist from books that may or may not have been written:

- *Lord Archibald "Feathers" Featherstone*
  Great game hunter and adventurer extraordinaire (just ask him). Brave, foolhardy, and frequently intoxicated.

- *Lady Clankington*
  Headmistress of Lady Clankington's Dueling Academy and Home for Wayward Girls and Young Offenders.

- *Professor Isadora Maelstromme*
  Genius inventor most noted for the helium expander. Resident inventor, building 5, Grimmelore Manor. Likes tea and has a wicked right hook.

- *Capt. Magnus Maelstromme*
  Brother-in-law to Isadora Maelstromme. Shady dealings of an unknown disposition. Frequently spotted at mysterious crash sites.

- *Dr. Adrian Grimmelore*
  Super genius. Reclusive millionaire. Mad, mad, mad scientist.

- *Ms. Adelaide Grayson*
  Major, Her Majesty's Aero-Forces, Ret. Spectacular pilot. Known to take her tea without cream and her scones unbuttered. She is well suited to survival in hostile foreign climes.

- *Espionagent Virgil J. Knockov*
  Spy and saboteur. An unsavory chap about whom little is known except that he always cheats at cricket.

- *Busty von Schluttentrappen*
  Spy and femme fatale works as an independent agent, yet has been known to be on the payroll from time to time of Herr Knockov.

- *Mr. Cecile F. Butterknuckles*
  Toady and henchman. Truly epitomized by the term "toady" because it describes him in both aspect and mannerisms. A ghastly chap, even in good light.

- *Captain Seamus Harper*
  B.O.R.G. 709th, veteran of the Colonial Wars and recipient of the Distinguished Imperial Cross for conspicuous acts of blind courage.

- *Colonel I.M. Havelleft*
  Britannic Officers Reconnaissance Group, Command Corps.

- *Sergeant Armstrong*
  Aeronautical boarding crew instructor and part-time motivational speaker.

- *Major Smythe*
  Hapless unwilling and unwitting time traveler often seen wearing a red shirt.

- *Ensign Ernest Lloyd*
  H.M. Geographical Positioning Service (air pirate and skywayman).

◦ *Air Admiral Horatio Lord Nimitz*
Well, what's to say…he's an admiral.

◦ *Chief Inspector General*
Sir Cockburn Taylor-Warres
Primarily responsible for defending the British Isles against marauding skywaymen.

◦ *Remy-Martin Courvoisier*
Infamous French skywayman plaguing the British coastline, who somehow speaks with an English accent.

◦ *Col. S.J. Waiselthorpe*
Reviled author of *Aero-Naval Boarding Actions Handbook.*

◦ *Old Lord Floggenhall*
Is dead.

For the Brilliant but Mathematically Challenged Mad Scientist

Metric to U.S. Conversions (Length)

1 millimeter	=	0.03937 inch
1 centimeter	=	0.3937 inch
1 meter	=	39.37 inches
1 meter	=	3.281 feet
1 meter	=	1.0936 yards

U.S. to Metric Conversions (Length)

1 inch	=	25.4 millimeters
1 inch	=	2.54 centimeters
1 foot	=	304.8 millimeters
1 foot	=	30.48 centimeters
1 yard	=	914.4 millimeters
1 yard	=	91.44 centimeters

Conversion Chart: Fraction/Decimal/Millimeter

factor 1/25.4 – 4 places (1 inch = 25.4 mm exactly)

Fraction	Decimal	mm	Fraction	Decimal	mm	Fraction	Decimal	mm
1/64	0.0156	0.3969	1 1/64	1.0156	25.7969	2 1/64	2.0156	51.1969
1/32	0.0313	0.7938	1 1/32	1.0313	26.1938	2 1/32	2.0313	51.5938
3/64	0.0469	1.1906	1 3/64	1.0469	26.5906	2 3/64	2.0469	51.9906
1/16	0.0625	1.5875	1 1/16	1.0625	26.9875	2 1/16	2.0625	52.3875

(continued on next page)

Fraction	Decimal	mm	Fraction	Decimal	mm	Fraction	Decimal	mm
5/64	0.0781	1.9844	1 5/64	1.0781	27.3844	2 5/64	2.0781	52.7844
3/32	0.0938	2.3813	1 3/32	1.0938	27.7813	2 3/32	2.0938	53.1813
7/64	0.1094	2.7781	1 7/64	1.1094	28.1781	2 7/64	2.1094	53.5781
1/8	**0.1250**	**3.1750**	**1 1/8**	**1.1250**	**28.5750**	**2 1/8**	**2.1250**	**53.9750**
9/64	0.1406	3.5719	1 9/64	1.1406	28.9719	2 9/64	2.1406	54.3719
5/32	0.1563	3.9688	1 5/32	1.1563	29.3688	2 5/32	2.1563	54.7688
11/64	0.1719	4.3656	1 11/64	1.1719	29.7656	2 11/64	2.1719	55.1656
3/16	0.1875	4.7625	1 3/16	1.1875	30.1625	2 3/16	2.1875	55.5625
13/64	0.2031	5.1594	1 13/64	1.2031	30.5594	2 13/64	2.2031	55.9594
7/32	0.2188	5.5563	1 7/32	1.2188	30.9563	2 7/32	2.2188	56.3563
15/64	0.2344	5.9531	1 15/64	1.2344	31.3531	2 15/64	2.2344	56.7531
1/4	**0.2500**	**6.3500**	**1 1/4**	**1.2500**	**31.7500**	**2 1/4**	**2.2500**	**57.1500**
17/64	0.2656	6.7469	1 17/64	1.2656	32.1469	2 17/64	2.2656	57.5469
9/32	0.2813	7.1438	1 9/32	1.2813	32.5438	2 9/32	2.2813	57.9438
19/64	0.2969	7.5406	1 19/64	1.2969	32.9406	2 19/64	2.2969	58.3406
5/16	0.3125	7.9375	1 5/16	1.3125	33.3375	2 5/16	2.3125	58.7375
21/64	0.3281	8.3344	1 21/64	1.3281	33.7344	2 21/64	2.3281	59.1344
11/32	0.3438	8.7313	1 11/32	1.3438	34.1313	2 11/32	2.3438	59.5313
23/64	0.3594	9.1281	1 23/64	1.3594	34.5281	2 23/64	2.3594	59.9281
3/8	**0.3750**	**9.5250**	**1 3/8**	**1.3750**	**34.9250**	**2 3/8**	**2.3750**	**60.3250**
25/64	0.3906	9.9219	1 25/64	1.3906	35.3219	2 25/64	2.3906	60.7219
13/32	0.4063	10.3188	1 13/32	1.4063	35.7188	2 13/32	2.4063	61.1188
27/64	0.4219	10.7156	1 27/64	1.4219	36.1156	2 27/64	2.4219	61.5156
7/16	0.4375	11.1125	1 7/16	1.4375	36.5125	2 7/16	2.4375	61.9125
29/64	0.4531	11.5094	1 29/64	1.4531	36.9094	2 29/64	2.4531	62.3094
15/32	0.4688	11.9063	1 15/32	1.4688	37.3063	2 15/32	2.4688	62.7063
31/64	0.4844	12.3031	1 31/64	1.4844	37.7031	2 31/64	2.4844	63.1031
1/2	**0.5000**	**12.7000**	**1 1/2**	**1.5000**	**38.1000**	**2 1/2**	**2.5000**	**63.5000**
33/64	0.5156	13.0969	1 33/64	1.5156	38.4969	2 33/64	2.5156	63.8969
17/32	0.5313	13.4938	1 17/32	1.5313	38.8938	2 17/32	2.5313	64.2938

Fraction	Decimal	mm	Fraction	Decimal	mm	Fraction	Decimal	mm
35/64	0.5469	13.8906	1 35/64	1.5469	39.2906	2 35/64	2.5469	64.6906
9/16	0.5625	14.2875	1 9/16	1.5625	39.6875	2 9/16	2.5625	65.0875
37/64	0.5781	14.6844	1 37/64	1.5781	40.0844	2 37/64	2.5781	65.4844
19/32	0.5938	15.0813	1 19/32	1.5938	40.4813	2 19/32	2.5938	65.8813
39/64	0.6094	15.4781	1 39/64	1.6094	40.8781	2 39/64	2.6094	66.2781
5/8	**0.6250**	**15.8750**	**1 5/8**	**1.6250**	**41.2750**	**2 5/8**	**2.6250**	**66.6750**
41/64	0.6406	16.2719	1 41/64	1.6406	41.6719	2 41/64	2.6406	67.0719
21/32	0.6563	16.6688	1 21/32	1.6563	42.0688	2 21/32	2.6563	67.4688
43/64	0.6719	17.0656	1 43/64	1.6719	42.4656	2 43/64	2.6719	67.8656
11/16	0.6875	17.4625	1 11/16	1.6875	42.8625	2 11/16	2.6875	68.2625
45/64	0.7031	17.8594	1 45/64	1.7031	43.2594	2 45/64	2.7031	68.6594
23/32	0.7188	18.2563	1 23/32	1.7188	43.6563	2 23/32	2.7188	69.0563
47/64	0.7344	18.6531	1 47/64	1.7344	44.0531	2 47/64	2.7344	69.4531
3/4	**0.7500**	**19.0500**	**1 3/4**	**1.7500**	**44.4500**	**2 3/4**	**2.7500**	**69.8500**
49/64	0.7656	19.4469	1 49/64	1.7656	44.8469	2 49/64	2.7656	70.2469
25/32	0.7813	19.8438	1 25/32	1.7813	45.2438	2 25/32	2.7813	70.6438
51/64	0.7969	20.2406	1 51/64	1.7969	45.6406	2 51/64	2.7969	71.0406
13/16	0.8125	20.6375	1 13/16	1.8125	46.0375	2 13/16	2.8125	71.4375
53/64	0.8281	21.0344	1 53/64	1.8281	46.4344	2 53/64	2.8281	71.8344
27/32	0.8438	21.4313	1 27/32	1.8438	46.8313	2 27/32	2.8438	72.2313
55/64	0.8594	21.8281	1 55/64	1.8594	47.2281	2 55/64	2.8594	72.6281
7/8	**0.8750**	**22.2250**	**1 7/8**	**1.8750**	**47.6250**	**2 7/8**	**2.8750**	**73.0250**
57/64	0.8906	22.6219	1 57/64	1.8906	48.0219	2 57/64	2.8906	73.4219
29/32	0.9063	23.0188	1 29/32	1.9063	48.4188	2 29/32	2.9063	73.8188
59/64	0.9219	23.4156	1 59/64	1.9219	48.8156	2 59/64	2.9219	74.2156
15/16	0.9375	23.8125	1 15/16	1.9375	49.2125	2 15/16	2.9375	74.6125
61/64	0.9531	24.2094	1 61/64	1.9531	49.6094	2 61/64	2.9531	75.0094
31/32	0.9688	24.6063	1 31/32	1.9688	50.0063	2 31/32	2.9688	75.4063
63/64	0.9844	25.0031	1 63/64	1.9844	50.4031	2 63/64	2.9844	75.8031
1	**1.0000**	**25.4000**	**2**	**2.0000**	**50.8000**	**3**	**3.0000**	**76.2000**

Final Thoughts

When I was asked to write this book, I hesitated at first. Not simply because writing a book is a heck of a lot of work, but, more specifically, I questioned the wisdom of giving away designs and techniques that I had spent so many years creating through a process of trial, error, cursing, more error, and hard work. In the end, I presumably make these things for a living. However, I have found, through experience, that each time I create a new design, it is inevitably co-opted by unscrupulous copycats who rarely seem to come up with any of their own designs. So, I thought, why not? If my designs are going to be copied, then why not provide everyone with instructions on how to create them? In doing so, it forces all of us designers to sort of "up our game." We all now have to come up with bigger, better, and badder variations and ingenious contraptions.

By now I am hoping that you will have managed to osmose the fundamental principles that underlie these sorts of Steampunk creations. There are no hard and fast rules, but there are certainly design and construction criteria that differentiate quality Steampunk gadgets and gizmos from cheap or hastily made tat. Now that you are familiar with the design and creation process, I encourage you to take these principles and develop your own ideas.

If you see me out and about at some event, convention, or gathering (assuming, of course, I manage to escape the demands of the workshop and laboratory), then please come up and talk to me. Ask me questions if you have them. Show me what you have made. Send us your pictures or tag us in online picture posts. We are all, after all, at the vanguard of this burgeoning aesthetic and subcultural movement, we are all forging future antiques, and we want to see where your imaginations lead.

Patterns from Chapter 5

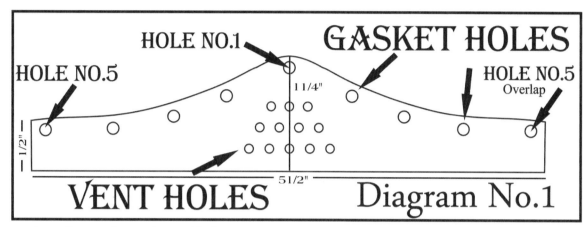

Figure 5-1 *Pattern for ocular and holes*

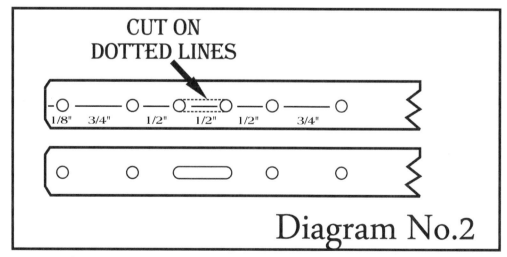

Figure 5-2 *Pattern for 1/2-inch strap buckle placement*

Index